Historic Industrial Scenes

SHEFFIELD STEEL

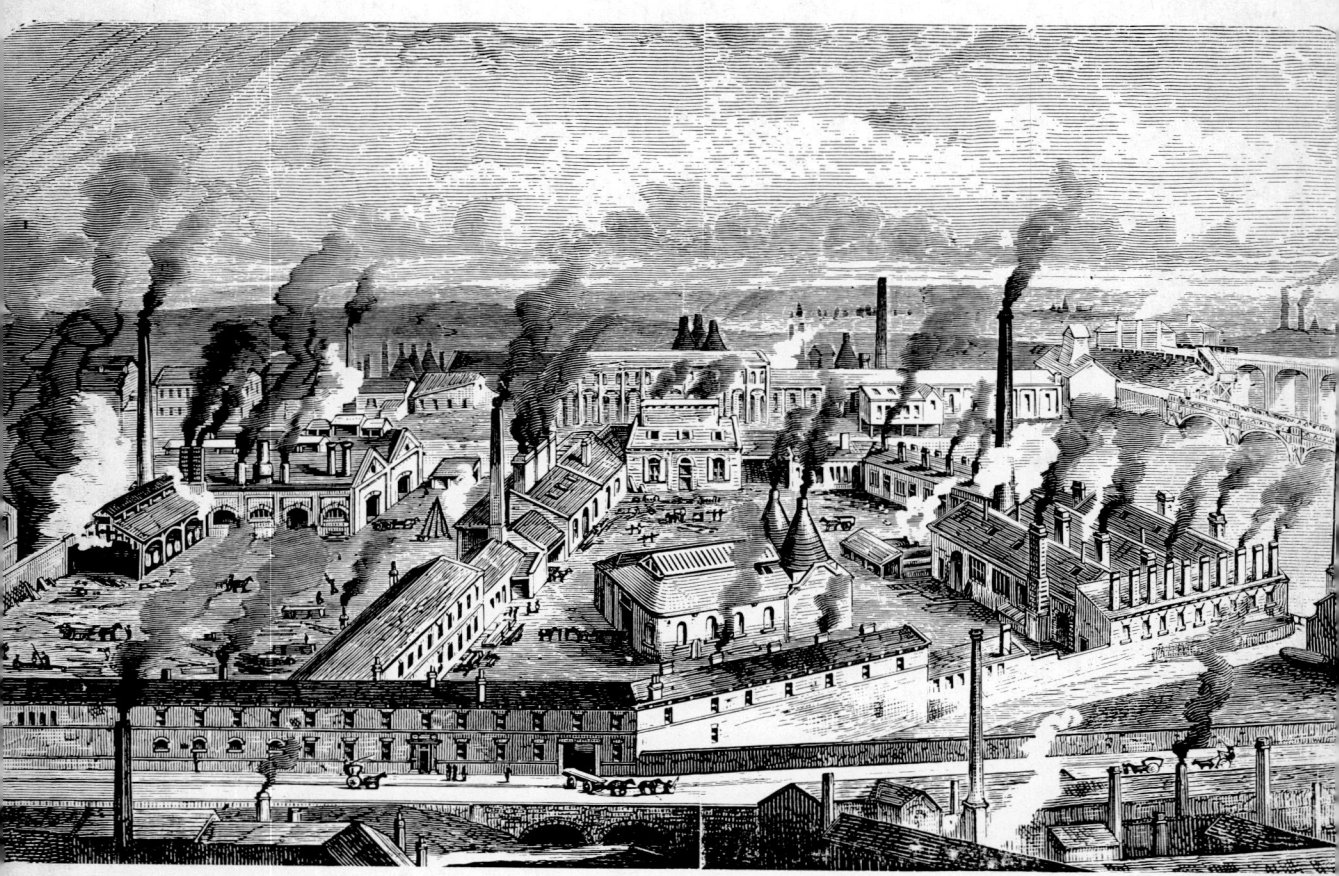

Scotia Steel Works: Messrs Thomas Jowitt and Sons. This illustration shows a typical medium sized steel works of the last half of the nineteenth century, built in Attercliffe Road in 1864. It is particularly interesting, however, in that it is taken directly from a woodcut block which was produced for an illustration in the *Illustrated Guide to Sheffield*, published in 1879. (Block kindly loaned by Frank Smith, Esq).

Historic Industrial Scenes

SHEFFIELD STEEL

K. C. Barraclough

MOORLAND PUBLISHING COMPANY

ISBN 0903485 31 1

© K. C. Barraclough 1976

COPYRIGHT NOTICE

All rights reserved. No part of this publication may be reproduced, stored in a retrieval system, or transmitted, in any form or by any means, electronic, mechanical, photocopying, recording or otherwise, without the prior permission of Moorland Publishing Company.

Printed in Great Britain by
Wood Mitchell & Co Ltd, Stoke on Trent

For the Publishers
Moorland Publishing Company
The Market Place, Hartington,
Buxton, Derbys, SK17 0AL

Contents

	Page
Acknowledgements	6
Foreword by W. G. Ibberson	7
Sheffield Steel	
1 The Background	8
2 The Beginnings	9
3 The Cutlers' Organisations	9
4 Production of Steel	10
5 Communications	11
6 Cementation Steel	12
7 Crucible Steel	12
8 The Bessemer Process	14
9 Open Hearth and Basic Steel	14
10 Forging, Rolling and Further Treatment of Steel	15
11 Postscript	15
Old Sheffield	16
The Cementation Process	22
The Supply of Iron	30
Crucible Steel Melting	36
Sheffield Steelworks	53
Later Steelmaking Processes	
Bessemer Steel	73
Open Hearth Steel	74
The Forging of Steel	79
Rolling of Steel	85
Heavy Engineering	91
The Lighter Trades	93
The Remaining Heritage	106
Bibliography	110
Index	111

Acknowledgments

The collection of the material for a volume of this type is inevitably the fruit of contact with many people over a period of several years, without whose assistance the task would not have been possible. While it should be self-evident that the true industrial archaeologist preserves complete records of what he handles, it has to be admitted that the origin of certain information is not clearly known to me at this present time and it is, therefore, difficult to give acknowledgement in all cases; should anyone feel that due credit has not been given, I would crave his indulgence and assure him that past help has been greatly appreciated and specific omission here stems from haziness of recollection rather than from any discourtesy.

Having said this, there are those who have been directly connected with this project in one way or another and whose assistance is recorded with gratitude. In particular, I wish to thank Dr D. Hardwick, Chairman and Managing Director of Firth Brown Limited, who not only granted permission for the reproduction of many items from the firm's archive collection of photographs but granted full use of the Firth Brown Photographic Department for the production of the illustrations; sincere thanks are also rendered to the staff of this department for their tolerance, as well as their efficiency. In connection with the preparation of the text, I have drawn considerably on an unpublished manuscript, written by Dr Allen McPhee in 1939 as part of a proposed 'History of Sheffield Trades', for the early beginnings of industry in Sheffield; this work, entitled *The Growth of the Cutlery and Allied Trades to 1814,* deserves to be better known and I gladly acknowledge my indebtedness to it. With regard to the several references to the Swedish travel diaries, I have to thank my very good friend Torsten Berg for unravelling the passages in the various manuscripts which have specific reference to our subject. W. G. Ibberson kindly made several useful suggestions, which were gladly accepted and Norman L. Clay kindly and painstakingly read the whole of the draft script, with great benefit to the final text; the invaluable assistance of both these gentlemen is gratefully acknowledged. Any errors which remain, whether factual or matters of interpretation, and any omissions, I must claim as my responsibility; I would, in fact, welcome any comments which could help in making the overall picture more clear.

Among those who assisted with the provision of ideas for the illustrations I would record the kindness of Charles Blick of BSC Special Steels Division, J. W. Silvester of the Sheffield City Museum, Martin Olive of the Local History Section of the Sheffield City Library, S. Boswell of Davy and United and Messrs Richard T. Doncaster, Kenneth W. Hawley and Frazer Wright.

With regard to the illustrations themselves, acknowledgement is due to Firth Brown (7, 21, 22, 23, 39, 49, 62, 64, 65, 66, 67, 103, 104, 105); Sheffield City Library (1, 2, 4, 70, 72, 73, 74, 79, 81); Graves Art Gallery (5); Beamish Museum (19); J. O. Vessey (26, 43, 44); R. T. Doncaster (27, 42, 155, 156, 157); Sheffield City Museums (6, 20, 158, 159, 160); British Steel Corporation (86, 92, 93, 94, 95, 96, 97, 98, 99, 100, 101, 117, 118, 119, 120, 124, 125, 126); Davy United (88, 127, 128). Other sources are as follows: *Annales des Mines,* 1843 (8, 32); W. Jessop, *Visit to a Sheffield Steelworks,* 1913 (9, 10, 24, 34, 35, 36, 37, 38, 40, 41, 51, 53, 56, 57, 77, 102, 107, 115); *Sheffield and Rotherham Illustrated,* 1897 (11, 46, 90, 116, 136, 138, 142, 144); G. Jars, *Voyages Metallurgiques,* 1774 (12); H. Seebohm, *On the Manufacture of Cast Steel,* 1869 (28); *British Steelmaker* (29, 58); *Med Hammare och Fackla,* 1932 (30); G. Broling, *Resa i England aren 1797, 1798 och 1799,* (31); *Metal Progress,* 1940 (33); *Huntsman's Catalogue,* c1930 (45, 50, 63); Carnegie and Gladwin, *Liquid Steel . . . ,* 1913 (47); Thomas Marshall and Co, *Calendar* c1942 (48); *Illustrated London News* (52, 54, 55, 91, 122); J. D. Scott, *Vickers – A History* (59); S. Pollard, *Three Centuries. . .* (60); J. Horsfall, *The Iron masters of Penns* (61); *The Sheffield Illustrated Guide,* 1862 (75, 83, 84, 85, 106, 114), and 1879 (68, 76, 78, 80, 82, 87, 89, 121, 130, 131, 137); Cammell's Brochure, 1898 (69, 71); J. Holland, *Sheffield and its Neighbourhood,* 1865 (108, 110); *Engineering* (109) Vickers, Son and Maxim, Brochure, 1898 (112, 123); John Spencer and Co (Newburn on Tyne), Centenary Brochure, 1910 (113); Arthur Lees, Brochure (date unknown) (129); Ward and Payne's *Sheffield List,* 1911 (133, 134, 139, 140, 141); William Marples *Sheffield List,* 1909 (135, 145, 146, 147, 148, 149, 150, 151, 152, 153, 154); Hunter's *Hallamshire* (3). Figs 13 to 18 are the author's copyright; the origin of Figs 25, 132 and 143 has not been traced. Thanks are also due to Frank Smith for the loan of the old woodcut block used for the frontispiece.

Finally I would like to record my gratitude to my wife, not only for her helpful suggestions and assistance with correction of the script, but for her forbearance over the period when the house was littered with photographs and pieces of paper.

K. C. B.

Foreword

It is a great pleasure to me to welcome the publication of *Sheffield Steel*, a survey which deals concisely with the historical development of steelmaking and explains clearly the evolution of the processes. The numerous illustrations have never before been presented together. They portray a most interesting record of the changing face of industrial Sheffield up to 1900. The author's captions reveal details which can so easily be overlooked and his explanations of them will promote a deeper understanding for many, if not all, of his readers. He has the gift of being able to describe technical matters so that people who are neither metallurgists nor engineers can readily understand them. His daily life is closely concerned with the making of Sheffield's famous special steels by the most modern methods, yet one of his main interests is industrial archaeology, which occupies much of his spare time.

In the archives of the Company of Cutlers in Hallamshire there are unique day-to-day records of the operation of a cementation furnace in which the Company made steel for the cutlery trade in the eighteenth century. For many years nobody had interpreted the quaint and laconic book entries. At my request, Mr Barraclough studied them and produced a most interesting account, which revealed how it was run, the production it achieved, the source of its raw materials and by what route they came. Very few people could have done this and I consider that we in Sheffield are fortunate to have such a man in our midst, who is willing to use his knowledge and his experience to place before us, especially in this book, so many interesting facts, which enable us to appreciate more deeply than hitherto the great accomplishments of our forebears.

He is a member of the South Yorkshire Trades Historical Trust, of which I am the Chairman, and, with characteristic generosity he has offered to give to the Trust all the royalties from the sale of *Sheffield Steel*. The members of the Trust tender to him their profound thanks.

The South Yorkshire Trades Historical Trust is a company limited by guarantee, incorporated in 1969, in order 'to protect and conserve industrial buildings and monuments of historical or architectural importance for the benefit of the public in general and the inhabitants of South Yorkshire in particular'. Its interests embrace the restoration of works and machinery, the preservation of manuscripts and pictures, in other words, everything considered to be part of our industrial heritage which should not be destroyed. The proceedings of the Trust are conducted by its Council of Management, whose main task at present is the restoration of Wortley Top Forge.

This is one of the oldest and most important ironworks in Britain, whose history is an integral part of the Industrial Revolution in this area and whose roots lie in the iron industry of many centuries ago. The present hammers and waterwheels are between 150 and 250 years old, but evidence is emerging of iron having been worked at Wortley in Saxon times. The members of the Trust receive no remuneration and they collaborate closely with members of the Sheffield Trades Historical Society, the South Yorkshire County Council and the Department of the Environment. The restoration of Wortley Top Forge is making satisfactory and steady progress. The forge will become an important place to visit, not once but many times, for there will always be something interesting to see.

Mr Barraclough has described in consummate fashion how the making of steel developed in Sheffield and its vicinity. I have expanded this foreword in order to show what it is, so dear to his heart, that he wishes to help by the sales of this book, namely the restoration and conservation of a fine old works of absorbing interest.

W. G. Ibberson

Sheffield Steel

1 The Background

Sheffield had a reputation for its cutlery long before it became the major steelmaking centre of Britain, a position which it undoubtedly held during the second and third quarters of the nineteenth century. Indeed, bearing in mind that the transition from one phase to another could occur over a period of time, some firms doing pioneer work while others capitalised on previous investments by working them longer, the history of Sheffield as far as steel is concerned falls into five periods which can be dated approximately as follows:

1 Up to around 1750 the area was a producer of iron but, in the main, an importer of steel. Imports up to the end of the seventeenth century would seem to have come mainly from the continent, but from 1700 or slightly later there was a change to supplies from around Newcastle.

2 From 1750 to 1800 occurred the development of steelmaking generally referred to as 'the Sheffield methods', namely the production of blister-steel in cementation furnaces and the melting of this steel in crucibles by the process invented locally by Benjamin Huntsman, to produce small ingots of the so-called 'cast steel' – steel had not been cast before!

3 From 1800 to 1865, based on these two methods with slight variations, a rapid growth of capacity from about 3,000 to 100,000 tons per annum provided a golden age (although those concerned with industrial pollution might well not agree with this epithet), where steelmaking not only became the major activity in the town, but was the basis of its prosperity and growth of population: from 10,000 in 1736 to 45,000 in 1801, with a threefold increase to 135,000 by 1851 and a threefold increase again to over 400,000 by 1901.

4 From 1865 to 1880 a mixed economy with bulk steelmaking processes (Bessemer and Open Hearth) introduced into the town, largely to meet a demand for steel rails. While these processes eventually were to produce far more steel than the old Sheffield methods, these latter continued to be important (indeed, 1873 was probably the zenith of their output, possibly of the order of 120,000 tons in the year) and continued alongside the newer methods.

5 From about 1880 onwards Sheffield became increasingly a centre for the production of special steel; in the first place, it was here the heavy items were made for ordnance, shipbuilding and general heavy engineering, to a large extent using the Open Hearth process to provide the steel. In addition, with the invention of the basic steelmaking processes by Gilchrist Thomas in 1879, steel production all over the world grew rapidly, and while this implied a shift of bulk steelmaking activities away from the Sheffield area and to the orefields, it also meant that more steel had to be machined and Sheffield's old steelmaking capacity was kept busy providing its traditional products of turning tools, saws and files. As time went on, there was a change in the types of material from the old carbon steel to alloy steel and here again Sheffield took up the challenge. While we are here only specifically interested up to around 1900, it is worth comment that, while methods have changed radically since that time, attitudes and patterns remain very similar and Sheffield steel is now quality-based special alloy-steel; the same could well have been stated quite truthfully in 1900.

Sheffield is situated on a bend in the River Don, at its confluence with the River Sheaf. Within a few miles of the centre of the town these two rivers receive as tributaries the Loxley, the Rivelin and the Porter and, a little to the east, the Blackburn and then the Rother. All these streams have their origins on the eastern slopes of the Pennines in an area with fairly high rainfall. They flow through well-wooded countryside and provide a constant flow of water throughout all the seasons. Their valley bottoms show that this was used as a source of power in days gone by, with their dams and remnants of watermills, which not only ground flour or fulled cloth, but drove bellows for furnaces, operated helve hammers or turned grinding wheels for the earlier industry of the area. The woods also provided fuel, the production of charcoal being a prime requisite for smelting and forging operations. It was only late in the eighteenth century that the value of 'pit-coal' was really accepted; this coupled with the development of the steam engine brought about the major industrial revolution – the freedom of industry from the tyranny of wood and water, as one writer so significantly put it. Sheffield, however, was well placed for the change, being situated on the rich South Yorkshire coalfield. Not only did this provide the fuel; it had previously given the iron ore for the charcoal blast furnaces. In view of what has just been said, it should be pointed out that the earlier workings in the coal measures were the bell pits of the iron-ore miners, who threw out the coal as unwanted rubbish (and this could be seen as black circles on the ground when the bell pit areas were levelled during recent open-cast coal mining operations). The coal measures also provided clays which were very suitable for the making of crucibles, the ganister rock, which was widely used in furnace construction and the various textures of stone from the Millstone Grit, used either for the provision of grinding wheels or the linings of blast furnaces or for constructing cementation furnace chests. Paradoxically, the terrain which provided all these advantages also proved a hindrance to the development of the town; the hilly countryside made communications difficult, so much so that isolated communities tended to specialise in one particular manufacture within their cottage industry; in addition, there tended to be specialisation in items which were easy to transport, but which had received considerable craftsmanship. Under these conditions, the cutlery industry flourished, while steelmaking and the heavier aspects of industry lagged behind until improved transport with the outside world was available, which became increasingly the case after about 1750.

2 The Beginnings

The Romans seem to have smelted iron at Templeborough. After the Dark Ages, there seems to have been renewed activity in late Anglo Saxon times; indeed, it is now claimed that ironworking sites in the area were destroyed by Canute and these must have re-established themselves fairly quickly, only to be 'wasted' by William the Conqueror half a century later. In the middle of the twelfth century the monks of Kirkstead had ironmining and smelting rights near Kimberworth and their 'grange' or storehouse still stands nearby at Thorpe Hesley. There was obviously a small-scale industry in the Sheffield area in 1268 since the Dowager Lady Furnival received by grant a third part of the monies realised from smithies in 'Schefeuld' and 'Halumpsire', from the sale of wood from the Park and from quarrying operations at 'Riveling'. The same locations were still providing revenue to the Lord of the Manor a century later. The Poll Tax records of 1379 show the operations within an area of a few miles from the town of some twenty-two smiths and five cutlers, and only a few years later Chaucer referred to a 'Sheffield thwytel'. The main cutlery-making centres at this time, however, were London (with about two hundred cutlers in 1450) and York (with about thirty); other cutlers worked in Beverley, Doncaster, Chester, Gloucester, Oxford and Salisbury. By 1500, however, York had declined in importance and Sheffield, together with Birmingham, was becoming recognised as a major producer of edge tools. This seems to have coincided with the growth of water-powered grinding operations, since between 1478 and 1552 there are records of such sites at Holbrook, Ecclesfield, Wisewood, Rotherham, Millsands, Lescar, Little Sheffield and Wadsley Bridge. In 1637 the Lord of the Manor leased no fewer than twenty-eight 'cutler wheels', which were very profitable to him and there were others in private hands at Ecclesfield, Norton and Eckington.

3 The Cutlers' Organisations

As the cutlers became more numerous and more prosperous, they sought to protect their craft and livelihood by recording their own trade marks and promulgating their trade regulations via the manorial court. John Parker, a scythesmith, bequeathed his mark to his son in 1552; the earliest known registration of such marks was, however, in 1554, when two knife-makers paid their annual rent of one penny and infringers were threatened with a fine of twenty shillings. We know that by 1568 some sixty marks had been registered. More important were the Cutlers' Ordinances of 1565, by which a minimum apprenticeship of seven years was established as the sole right of entry to the trade, stipulating two close-seasons of a fortnight in August and a month at Christmas and banning the making-up of materials supplied by outsiders or the supplying of outsiders with partly finished articles. These rules, enforced by fines in the manorial court, claiming to be in accordance with the ancient customs and ordinances, are of interest in suggesting that there were prior regulations which have not survived. They were revised in 1590: the close season in August became a month, a second apprentice could be taken, but only when the first had served six years of his term and entry to the trade could now be purchased for a sum of £5. The rules were slightly revised again in 1614 and by then 182 marks had been registered.

In 1617 the Lord of the Manor, the Earl of Shrewsbury, died without male heir and the manorial system of control of the cutlery trades broke down. A new system was needed and, after some abortive activity, an Act of Parliament in 1624 established 'The Company of Cutlers in Hallamshire', with rules very similar to the previous ordinances; while close-seasons disappeared, except for Lord's days and holy days, the minimum apprenticeship of seven years remained. A second apprentice could now be taken on when the first had served only five years, although this could be compounded by a fine; only masters, not journeymen, could take on apprentices and entry to the trade by purchase was now prohibited. At the time of the granting of the Charter, there were 498 masters (440 knife makers, 31 shear and sickle makers and 27 scissor makers). As time went on, other trades were admitted and of the 2,054 masters in 1682, in addition to the 1,562 knife makers, 137 shear and sickle makers and 284 scissor makers, we find 33 scythe makers had entered at a fee of £30 each, 21 file smiths at a fee of £20 each and 17 awl-blade smiths at £10 each.

4 Production of Steel

Our topic, however, is the story of Sheffield Steel. Truth to say, the bulk of the cutlery we have so far described seems to have been produced from imported steel, from Central Europe or from Spain. Local supplies were uncertain in quality. It is just possible that some steel was produced in the Rotherham area from about 1650 onwards, but the first definite reference to Sheffield steelmaking comes in 1692 when a Swedish visitor found that operations in this area were similar to those in Stourbridge (probably the main centre at this time), but implied they were on a much smaller scale. In 1709 we know that Samuel Shore was making steel in Sheffield; in 1720 another Swedish visitor mentioned two furnaces in the town, one belonging to Shore and the other to Perkins, with two other furnaces, somewhat larger, in Rotherham. In 1737, when Thomas Oughtibridge drew his 'prospect' of Sheffield and was careful to show 'the steel furnaces', there were still only two of them, both being what became known as 'cementation furnaces' or 'converting furnaces', their product being 'blister steel'.

To understand the techniques employed in the manufacture of steel it is necessary to go back to iron smelting. The Romans, and the monks of Kirkstead, built small bowl-shaped furnaces in which they mixed iron ore and charcoal and, having lit the charcoal, fanned the flames inside the furnace by the use of hand- or foot-operated bellows. Where the prevailing wind was strong they might also site their furnaces on the windward side of a hill so as to utilise the natural draught and this is the origin of the 'bole hills', as at Walkley. The effect of all this was to allow some of the oxide of iron, the chemical present in the iron ore, to react with some of the carbon, the main constituent of the charcoal, the remainder being burned to produce the necessary heat, and thus produce some metallic iron, together with a slag covering. This slag consisted of the remainder of the iron oxide together with the clayey and sandy materials present in the iron ore. The iron did not actually melt; it was a mass of small metallic particles, rather spongy in character, having the gaps between the metallic particles filled up with slag, and this mass was known as a 'bloom', the furnace being a 'bloomery'. On breaking down the front wall of the furnace, the bloom could be taken out, reheated in an open charcoal fire and then hammered to expel the bulk of the slag. In this way a bar of compact metal, soft and malleable, could be formed. This was bloomery iron, very similar to the later and better-known 'wrought iron'. If it was hammered out into strips, similar in form to knife-blades and these then reheated inside a glowing pile of charcoal, some of the carbon might well enter into the body of the iron; if so, it would be found that the material, if quenched from a red heat into a tank of cold water, would become hard – much harder than the original iron – and could be ground to make a cutting edge which would be reasonably durable. It had, in fact, by absorbing carbon, become steel of a sort. Steel, until late in the nineteenth century, was simply an alloy of iron containing between 0.5 per cent to 1.5 per cent carbon. After that time, other elements such as nickel, chromium, tungsten and so on were added as well as carbon and thus we entered the age of 'special steels' or 'alloy steels', the type of material for which Sheffield has been noted for the last fifty years or so.

As time went on, the bloomery increased in size. The small bellows were replaced by larger ones operated by water power, the bowl became a shaft and then something rather surprising happened. The temperature in the hearth increased from about 1,000°C in the small bloomery to over 1,200°C; at this temperature the iron began to absorb carbon very quickly from the charcoal and began to melt; the slag in the bottom part of the furnace also became fluid and covered the molten metal. Now instead of having to break down the furnace to extract the metal, if a small hole was made at the bottom the metal would run out and could be allowed to solidify in depressions made in a bed of sand in front of the furnace. The product was, however, nothing like that from the bloomery; if reheated and hammered it broke into pieces; on the other hand, if a complicated shape was made in the sand bed the metal would take up that shape. It was, in fact, 'cast iron'; it could be cast into rectangular blocks, which were known as 'pigs' or 'pig iron' and the furnace had become a 'blast furnace'. The first blast furnaces in the Sheffield area seem to have been at Wadsley and at Chapeltown, both built around 1600, using iron ore from the Tankersley coal seam, this being mined in bell pits, the coal being thrown away as unwanted!

Cast iron is a metal with about 4 per cent of carbon, some smaller amounts of silicon and manganese, the remainder being iron. To produce a forgeable iron from such a material it had to be remelted in a furnace known as a 'finery', under a bed of charcoal, and a blast of air blown in to burn out the carbon. As we have seen, the cast iron melts at about 1,200°C; as the carbon burns out, however, the melting point rises (until with only the 0.1 per cent carbon or so present in 'wrought iron' the melting point is around 1,500°C). Thus, as the process in the finery went on, the molten iron lost its carbon and became pasty; finally it could be removed as a semi-solid mixture consisting mainly of iron particles with some entrained slag – essentially similar to the 'bloom' from the bloomery – and this 'bloom', for so it was still termed, was hammered under a 'helve' (a hammer driven by water power) to consolidate it. It was then reheated again in a charcoal fire, this time in a furnace known as a 'chafery' and forged down to 'bar iron' or 'wrought iron' using a smaller helve hammer.

From what has been said of the carbon contents of the various materials it may have been seen that steel lies somewhere between wrought iron and cast iron in its content of this element. It is logical, therefore, to conclude that, could the operations in the finery be controlled and only part of the carbon from the cast iron burned out, steel could be produced. It is easy for us to say this; to the seventeenth-century ironmaster, however, it was all rather like witchcraft since he had no clear conception of the role played by carbon. Nevertheless a process had been developed in Austria whereby, after almost completing the finery process, some fresh cast-iron was introduced and worked into the metal; with a bit of luck the forged product would harden on quenching and thus they had managed to produce steel. Such was the material imported into Sheffield in the seventeenth century; it came in via the Rhine valley and therefore became known as 'German Steel' (or even 'Cullen Steel' since it passed through Cologne on its journey here).

The alternative method of making steel, already touched upon, depends on the introduction of a suitable amount of carbon into wrought iron. This is an extension of the principle of 'diffusion' of carbon from the charcoal fire into the iron knife-blade; such processes are more successful at higher temperatures and proceed further into the mass of the metal if the time of exposure is prolonged. But to prolong the life of a charcoal fire is difficult; if, on the other hand, the iron was bedded in charcoal in a closed vessel which could then be heated externally to the appropriate temperature, the heating could, if necessary, go on for days. There would be only one proviso: the heating must not be so fierce as to exceed the temperature at which cast iron would melt, since then the iron would absorb sufficient of the carbon from the charcoal to convert itself into a pool of liquid metal in the bottom of the container. Sometime just before 1600, someone, somewhere in Germany, appears to have worked this all out, without precisely knowing what he was doing. From there it spread westwards, and in 1613 an English patent for such a process was granted. The original patentees do not appear to have been particularly successful, although they did establish that the 'chests' containing the iron buried in charcoal could be heated externally by a pit-coal fire. This was the beginning of the 'cementation process' for the production of steel in this country. A number of years were to go by before it really became a feasible process, but it was carried out successful in Gloucestershire around 1635. The first contemporary description of the process comes from the same area in 1686; this account contains a very significant statement to the effect that the iron used was of Swedish or Spanish origin, since English iron was not suitable for making into steel. This was to prove to be the case for the next two hundred years or more, wherever the process was employed, and vast quantities of Swedish iron were imported into the Sheffield area in the nineteenth century—and also into the twentieth.

Sometime just before 1700 Ambrose Crowley, steelmaker of Stourbridge, moved his activities to the North East, finally settling in the valley of the Derwent, a tributary of the Tyne. This was to prove the major steelmaking area in the country for over fifty years. There is a fascinating story in the various travel diaries left behind by foreign visitors to this area in this period, the Crowley works being the magnet which drew them; it was the largest metal-working establishment in the whole of Europe. There is, in fact, very little remaining of these great works, but a few miles away, at Derwentcote, is one of the very few remaining cementation furnaces in the country. This is unique in that it is stone built, unlike the Sheffield furnaces which were invariably of brick; it dates back to 1740 and was quite a large furnace for its time. From the records still surviving from the Crowley works it is confirmed that their production of steel was confined to one particular grade of Swedish iron – this being known either as 'Dannemora' or 'Oregrund' iron, since it was made from Dannemora ore and the iron shipped out from the port of Oregrund.

5 Communications

The Newcastle area was uniquely favoured for such production: ample supplies of coal, easy access to a North Sea port, wooded countryside for the provision of charcoal, coal-measure clays and refractory sandstones. In contrast, however, at this time, communications in the Sheffield area were primitive and access to sea-borne imports extremely difficult and it seems to have been this alone which delayed the growth of steelmaking in South Yorkshire. There is no doubt that the roads were poor – so much so that carriage charges were higher in winter than in summer for the transport of pig iron from Bank Furnace to Wortley Forge in 1696. Earlier, in 1657, there had been complaints from those living in Attercliffe of their difficulty of attending the parish church in Sheffield two miles away, 'whereunto the road is often obstructed by floods', the Lord of the Manor's private road through the Park being apparently in common use from Civil War times until 1692. During this period, as for centuries previously, the wares of Sheffield had been taken on pack-horse trains to Bawtry, on the River Idle, and thence via the Trent and the Humber to the sea and to London and elsewhere along the coasts. The setting up of a network of turnpike roads, from 1756 onwards, eventually provided reasonable transport in most directions from the town. The main step in opening up the area, however, was the canalisation of the River Don. This began in 1727; by 1734 barges could go from Rotherham to Goole; by 1751 canal access to Tinsley was completed and arrangements made for the road from Tinsley to Sheffield to be well-maintained. The last canal stage (Tinsley to Sheffield) was a difficult one and was not completed until as late as 1819. But even without this, traffic in the latter part of the eighteenth century was considerable, some 13,000 tons of goods being moved by barge in a year. Moreover, instead of a charge of 30s (£1.50) per ton by packhorse for the journey to Doncaster or 15s (75p) per ton by cart, if the state of the road permitted, the charge by water was only one quarter as much, only some 5d (2p) per ton-mile. How much of this was iron import and cutlery export is not stated; we know, however, that it was estimated in 1802, when the extension of the canal was being promulgated, that the Sheffield steelmakers would utilise some 3,000 tons of Swedish bar iron a year, this being in addition to the amount used by the Walker concern at Masbrough.

6 Cementation Steel

Thus, in the latter part of the eighteenth century Sheffield began to make its name as a steel-producing region. The earliest furnaces in Sheffield of which we have a description date from 1767 and were capable of converting only 3 or 4 tons at a time and even in 1800 it seems likely that the Rotherham furnaces were still bigger than those in Sheffield. Over the next fifty years, however, well over two hundred furnaces were built in the Sheffield area, with capacities in some cases of as high as 40 tons; in 1843 Sheffield produced 90 per cent of the total British steel and almost 50 per cent of the European output. Such had its reputation become that visitors from the Continent came frequently to inspect the operations and it is from their reports that we get first-hand accounts.

What, then, was the process for the production of blister-steel in a converting (or cementation) furnace? With the drawings of Professor le Play, who visited Sheffield in 1842, to help us, we can note that the furnace could be recognised from outside by its conical chimney, from 35ft to 60ft high, and most of the representations of the old Sheffield works show one or more such structures. Inside the chimney were two chests (or 'pots'), made up from sandstone slabs or, more rarely, built from firebrick, set either side of a central flue, with subsidiary flues and chimneys providing the draught for the firegrate below so as to envelope the chests with flames. Into the chests went first of all a layer of charcoal, then a row of bars of Swedish iron, usually rectangular in section, $3\text{in} \times \frac{3}{8}\text{in}$ being quite a usual size, the bars running the length of the chests which could be up to 15ft. Then a further layer of charcoal went in on top of the iron, another layer of bars, and so on, until the level reached within 6in of the top. After a final layer of about 3in of charcoal, a thick pie-crust of 'wheelswarf', the debris from the bottom of the grinders' troughs which consisted of fine particles of sandstone mixed with steel grindings of a cement-like consistency. Then all the charging holes in the furnace were bricked up and sealed with clay before lighting the coal fire below. In about a day the chests were at a bright-red heat (about 1,100°C) and the firing was then continued for a further six to eight days. Then the fire was raked out and after about a week the interior of the furnace had cooled sufficiently for the doors to be opened up and a man to enter the furnace, break through the crust and pass out the bars to his mate outside, just as he had received them about three weeks earlier. The bars were originally smooth-surfaced and very tough and difficult to break; now they were covered in small blisters and could be fractured by a single blow with a hammer on an anvil – from iron they had been changed into 'blister steel'. By the appearance of the fracture the skilled workman could sort them into batches according to the amount of carbon they had absorbed.

They were of no immediate use as they were. They were, therefore, heated to a bright red and then either forged or rolled into bars. Sometimes the long bars would be broken into lengths of about 18in and a bundle of ten or twelve of these pieces, held together with strong wire, would be heated up and forged as a single piece into a bar, producing 'shear steel'. In this the various pieces welded together to give a more uniform material, but there were still alternate layers of high-carbon and low-carbon material; such a metal had a remarkable combination of flexibility and the capability of holding a sharp cutting edge, very suitable for the production of cutlery blades and other edge tools. To make it even better, the bar of shear steel, while still hot, could be cut in two and the two pieces again welded together under the hammer, thus producing the famous 'double-shear steel', renowned throughout the world as the supreme symbol of quality in Sheffield cutlery.

7 Crucible Steel

On the other hand, such steel was not universally suitable for all applications for which steel was required. Benjamin Huntsman, the Doncaster clockmaker, found it unreliable for springs and hit on the idea of making blister steel more uniform and more reliable by remelting it in small crucibles. His experiments took a number of years, but by 1751 he seems to have brought his method to a stage where he could undertake commercial production. Within ten years he was making up to 10 tons a year and his fame was such that foreign visitors were trying to gain entry to his 'works' – which were just a back-yard establishment with access through the house itself! There are, of course, legends that one of his rivals had stolen his secret by stealth, having disguised himself as a tramp seeking warmth and shelter for the night and thus gaining access to the melting shop. Be that as it may, Huntsman laid the real foundation of Sheffield's steel industry, his crucible steel or 'cast steel' being recognised world-wide as an improvement on any steel previously available. It was arguable that double-shear steel might be superior for cutlery blades due to its special qualities, but for almost any other use, for turning tools, chisels, razors, wire, sheet for the making of pen nibs, and many other applications, crucible steel reigned supreme throughout the nineteenth century.

The Huntsman process was very simple in concept: it involved the breaking up of blister-steel bar, putting the pieces into a crucible and melting the metal by burning coke around the covered crucible. To effect this, however, a furnace temperature of around 1,600°C was needed, much higher than had previously been used in any metalworking process and, indeed, higher than that needed for glass melting. This required a plentiful supply of air (hence the large cellar below the furnace), a strong draught (hence a tall chimney) and, above all, a crucible which would withstand the attack of the metal at this high temperature. As a result of individual experiments such crucibles were developed, from the small ones used by Huntsman himself, which held some 8-10lb of metal, to those used in common use in 1880, with a capacity of 60lb, or even 70lb. Most firms had their own 'recipes' for clay mixtures and all of them made their own crucibles on the premises. The 'potmaker' was a very important member of the team, treading the clay to the right consistency and moulding the crucibles, allowing them to dry slowly for two to three weeks on racks over the furnaces and finally bringing them up to a red heat overnight prior to introducing them into the furnace for the actual melting operation.

Crucible furnaces, like cementation furances, could also be recognised by their chimneys. These were some 30-40ft high, but set in batches of three to six flues in a single stack, forming a tall rectangular block above the furnace building. As a supply of blister steel was necessary for the remelting process, the 'integrated steelworks' of the first half of the nineteenth century would contain both the conical cementation furnace chimneys and the rectangular crucible furnace chimneys; smaller works, however, would generally buy in their blister steel, as they did at the Abbeydale Hamlet, where there was no cementation furnace.

The melting procedure consisted of setting a small coal fire on the cleaned furnace bars at six o'clock of a morning, putting the preheated crucible on its refractory-clay stand in the middle of the fire, placing a clay lid on the crucible, building coke around the crucible, placing the cover over the furnace hole and then firing it with full draught to bring it up to a brilliant white heat. Then the furnace cover was removed, the crucible lid moved sideways and the charging-funnel put into the crucible top. The previously-weighed charge was then carefully slid into the crucible through the funnel, the crucible lid replaced, the whole of the space around the crucible again filled up with coke, the furnace cover replaced and again a full draught applied. After two further charges of coke, at about hourly intervals, the charge would be molten; the melter from time to time would test the contents with an iron rod. When molten, another charge of coke and about one hour's further heating would bring about 'killing with fire' – in other words, the metal was brought into the right condition so that it would solidify without blowholes through the structure of the metal. The pot would then be pulled out from the furnace and placed on the floor, carried to the casting pit, the lid removed, the slag skimmed off and the metal poured, by hand, into the waiting ingot-mould, which with a 60lb charge would usually be some 3in square. Huntsman's original moulds seem to have been octagonal in section, about 2in across. We are so used to considering ingots as an essential part of steelmaking that it is hard to realise that Huntsman was the first to produce an ingot of steel, at least in Europe, just over two hundred years ago. (It is just possible that the Chinese anticipated him by a few hundred years, but this is not confirmed.)

By 1835, Sheffield was producing some 10,000 tons of steel a year. Then in 1838 the Sheffield to Rotherham railway was opened, the Wicker Station, the Sheffield terminus, being on a green-field area east of the town. Over the next twenty-five years the whole pattern changed, from almost a backyard industry to the East End pattern which we now know. Charles Cammell, Thomas Firth, John Brown and William Jessop all built their establishments on this open territory, with grid planning, either side of the railway; finally, Naylor, Vickers and Company built their massive River Don works in 1862. By 1873 at least 100,000 tons of crucible steel was being produced in the Sheffield region a year, a tenfold increase in just over thirty years.

8 The Bessemer Process

The scene was changing, however. In 1856 Bessemer announced his new process to the world. In truth, this proved a little premature, but within ten years the age of bulk-steel manufacture was firmly established and by 1870 more steel was being produced in Britain by the Bessemer process than in crucible pots. Sheffield steelmakers were slow to take up the new method; rightly so, since the cheaper Bessemer steel was more a replacement for wrought iron, being unsuitable for edge tools and the more critical applications. Nevertheless, eventually John Brown and Charles Cammell took up the Bessemer process – they seem to have been ready to try their hands at most things, having made wrought iron, and steel, by the puddling process from about 1857 onwards – and both built up a very profitable trade in steel rails until the early 1880s. On the other hand, for anything which required 'quality' the crucible process still reigned supreme and there are reports of a number of really large ingots – up to 25 tons in weight – being made by pouring the contents of five or six hundred crucibles into a tundish (a large cylindrical vessel, lined with firebrick, with a nozzle set in the bottom, capable of holding a quantity of liquid steel and allowing a constant stream to run out into an ingot mould below) and thus keeping the metal flowing for periods of over half an hour; needless to say, such an ingot was likely to be forged into a gun barrel, since money was never stinted on articles of destruction! And what rivalry there was between the Sheffield steelworks; while one perfected guns and shells, others made better and better armour-plate to resist them! Some firms even made both.

9 Open Hearth and Basic Steel

There was another steelmaking method being perfected at this time; this was the Siemens (or Open Hearth) process. A slower method than Bessemer's, it was more easily controlled and for quality steel gradually became the accepted method for the provision of ordnance and for engineering forgings for ship shafting, power transmission and the like. By 1900, when our survey closes, it had largely taken over in Sheffield from other processes, except for the remaining crucible steel produced for turning tools and the like. Both the Bessemer and Siemens processes have been so well described on many occasions that space will not be occupied here except to point out that both depended on the burning out of carbon from pig iron – in the Bessemer process liquid metal from the blast furnace had air blown through it, while in the Siemens process cast iron was melted on an open hearth, usually together with a certain amount of scrap, under a slag to which iron ore was added, this having the effect of oxidising or 'boiling out' the carbon from the metal. The Bessemer furnace could provide up to 25 tons of steel in half an hour; the Siemens furnace ultimately reached a capacity of 100 tons or more, but one 'heat' would take 10 to 12 hours.

Both these processes were altered in scope considerably by the invention of the 'basic' process by Gilchrist Thomas in 1879. This permitted the elimination of phosphorus, an element which is deleterious in steel when present in quantities over about 0.025 per cent, causing it to be brittle. The higher the carbon content the more problem there was with this unwanted element. So far, it will be remembered, the Sheffield steelmakers had made their steel almost exclusively from Swedish iron, low in both sulphur and phosphorus, and they continued in this insistence as long as these old processes survived; home-produced iron was too impure. Up to 1879 the only way of controlling the phosphorus content of Bessemer and Siemens steel was to use low-phosphorus pig iron, that is the iron ore fed to the blast furnace must be low in phosphorus. Britain was fortunate in having plentiful supplies of such ore in the haematites of the Furness district of Lancashire and this explains the rise of ironmaking and steelmaking in the district around Barrow in the twenty years from 1860 to 1880. On the continent, to the contrary, supplies of low-phosphorus ores were scarce, except in Spain (and the Spanish minerals were mainly bought up by British firms to supplement their own supplies!). Thus with these 'acid' processes (this implies furnaces lined with siliceous materials) Britain maintained its advantage in steelmaking. After 1879, however, largely as a result of a British invention, the odds were loaded against Britain. We had large supplies of phosphoric iron-ores and these now became usable; thus there was a shift of the industry to Cleveland, Lincolnshire and Northamptonshire. On the other hand, the Germans and the French and, in particular, the Americans could now tap immense resources of ore which had previously been useless and by 1890 American production of steel had outstripped the British and we were beaten into third place by Germany before the end of the century. The effect of all this on Sheffield was to change it to a centre producing special steels; the greater quantity of general-purpose bulk steel that was produced elsewhere in the country meant that more tools would be required to shape it and the old Sheffield methods were still the only ones capable of providing steel of the desired quality. In due course, the skilled Sheffield workers turned their attention to alloy steels and many of these – including what is probably the most famous of all, stainless steel – were invented and perfected in Sheffield. Sheffield still retains its reputation for special steel; while it now provides only about 5 per cent of the total British steel tonnage, its sales have a value amounting to almost 50 per cent of the total.

10 Forging, Rolling and Further Treatment of Steel

Apart from the actual steelmaking, something must be said of the operations carried out to make the steel suitable for the many uses to which it is put. We have touched on the use of water power for driving forging hammers and, later on, rolling mills. Such items persisted in the older works late into the nineteenth century. Steam power, however, after its tentative introduction as early as 1787, gradually took over and was installed from the outset in the new works erected in the east end of the town from 1845 onwards. Nasmyth invented his steam hammer in 1842 and by 1865 very large hammers, of 25 tons force, were being installed by John Brown and others. Shortly afterwards, hydraulic presses came in for forging the ever-larger ingots being made. The lighter side of the trade was handled by tilt hammers, the lineal descendents of the old water-driven helve-hammers. In these various ways could the forgings required by the engineering trades be provided. They still required heat treatment to develop their optimum properties and then had to be machined to the required shape and thus large machine-shops grew up in the main steelworks, generally driven by steam-powered shafting and belting.

On the rolling side, the mill train driven by a water wheel gave place to steam-powered types and the small rod and bar mills were supplemented by much larger 'cogging' mills, capable of rolling the larger ingots, up to three or four tons in weight, produced by the bulk steelmaking processes from about 1865 onwards. Large mills for the production of plates also came in about the same time; these eventually would have rolls up to 4ft in diameter with barrels of up to 6ft long, capable of producing plates weighing up to 20 tons.

Meanwhile the old 'lighter trades' still flourished in the town, largely operating on crucible steel. The cutlers, the makers of files, chisels, turning tools, machine tools of all kinds, scythes, sickles, scissors, razors and other edge tools had long traditions in this area and skilfully adapted themselves to the changing conditions, as they have continued to do to this day.

11 Postscript

There is so much to be told up to the close of the nineteenth century that further extension would be too large a task. There are some cases, however, where the illustrations which follow have been taken from the twentieth century. This has merely been done, however, where the necessary illustrations from an earlier source have been lacking.

This volume is in reality a tribute to the former craftsmen of Sheffield, many of whose descendents still ply the same skills and still have the same pride in the achievements, past, present and future, of this Yorkshire town. It is set, like Rome, on seven hills, separated by streams whose valleys are rich in industrial history and whose waters combine to swell the River Don, which has served to transport Sheffield's products, made from the highest-quality steel and enhanced in value by the skill of the craftsmanship applied to it, to the ends of the earth. There are many omissions; the passing of time has dealt more kindly with some records than with others and at this stage the loss of some items can only be noted with regret. In addition, no one author can do full justice to such a project since he can only report those things which have come to his notice, however diligent his research has been. It is hoped, however, that the reader will find in this offering an acceptable account of some of the mysteries of steelmaking, together with an interesting collection of old prints and photographs.

Old Sheffield

1 Gosling's Plan of Sheffield, 1736. This is the oldest-known plan of Sheffield. Though in 1736 it was only a small town (population under 10,000), apart from cutlers' workshops there were larger units – a tilt hammer (bottom centre) and the Wicker Wheel (in the bend of the River 'Dunn', below the title cartouche).

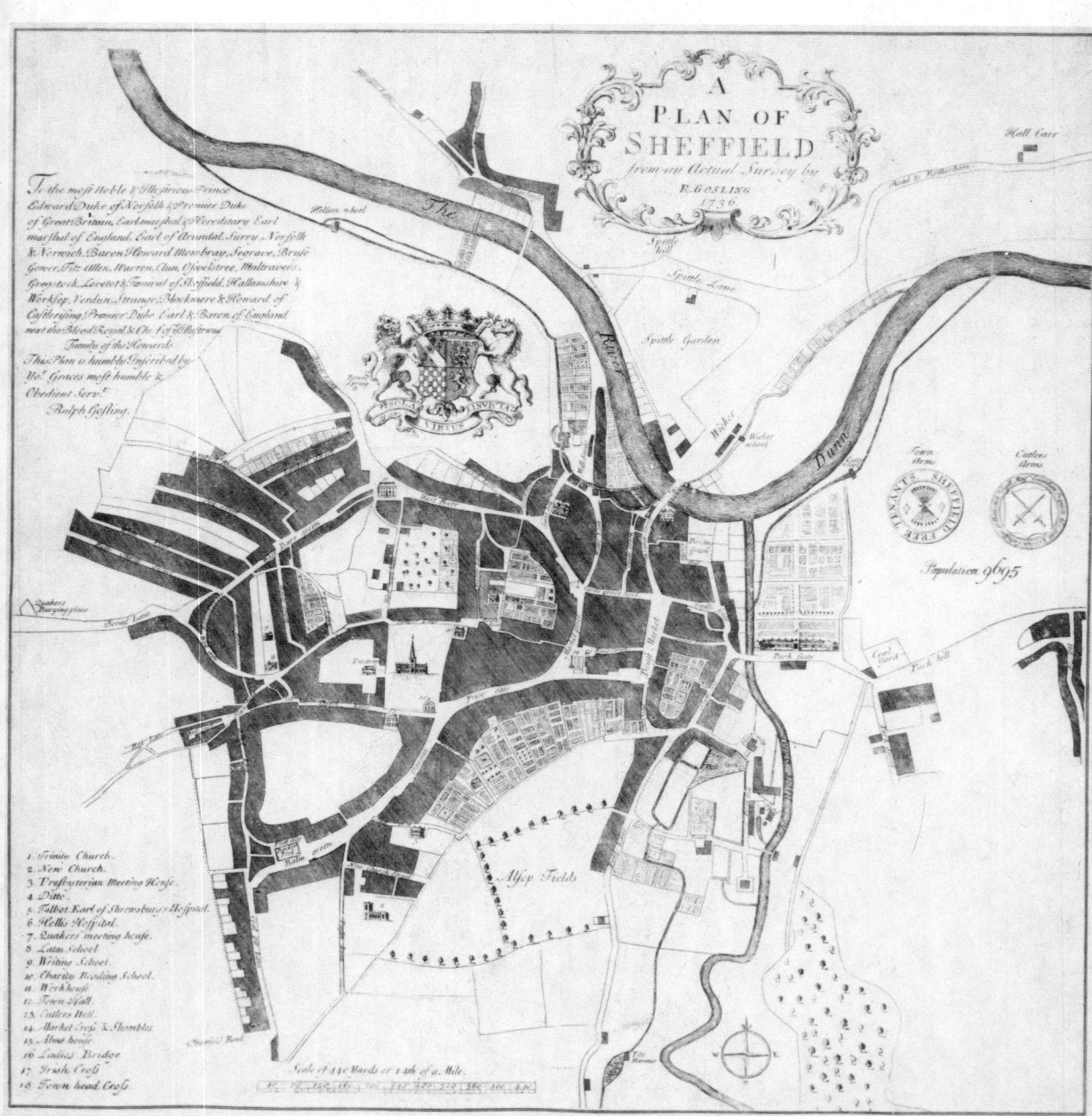

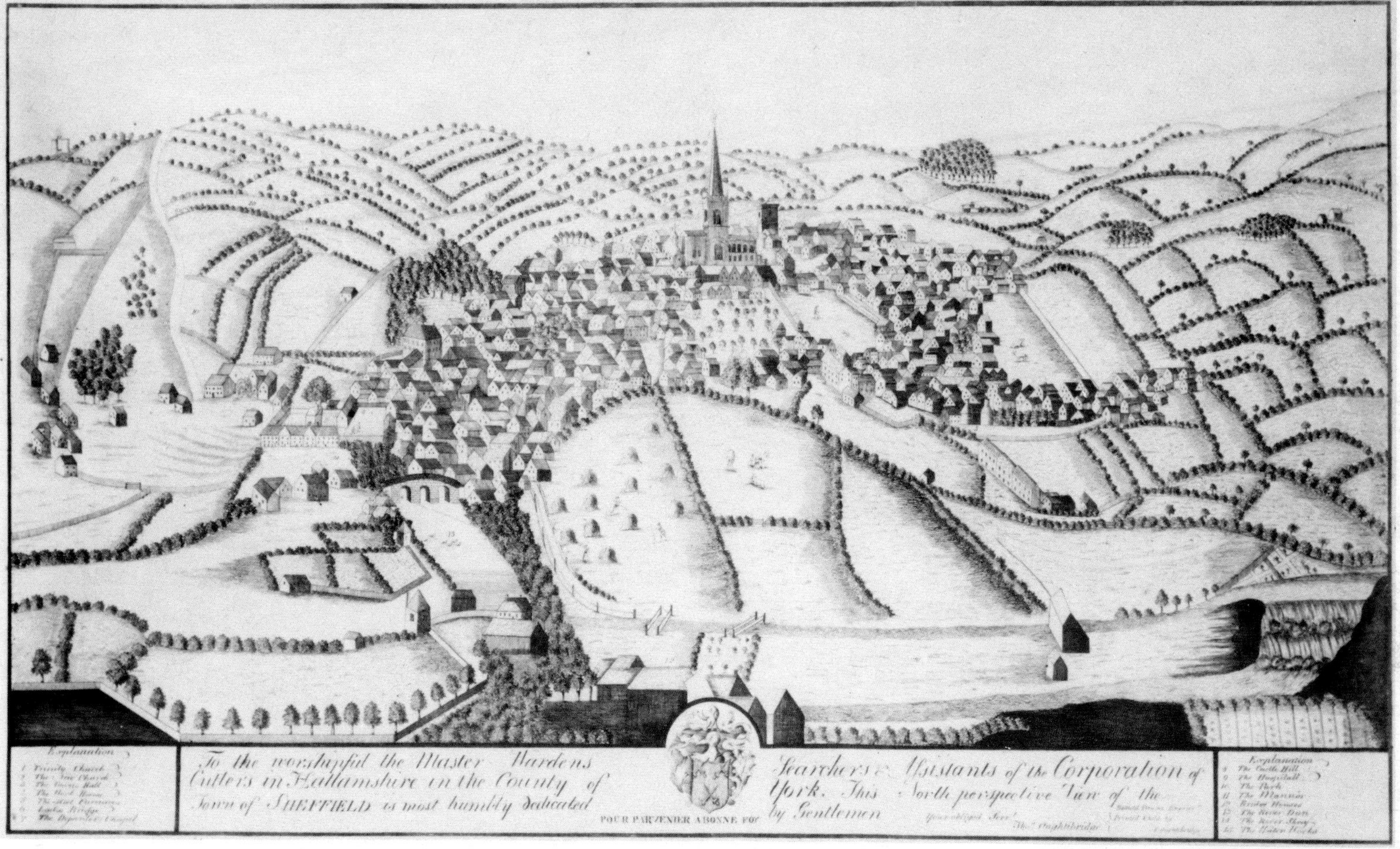

2 North Perspective of the Town by Thomas Oughtibridge, 1737. Virtually contemporary with Gosling's Plan, the main interest to us is '5 – The Steel Furnaces'; these are to be found directly below the copse on the horizon to the right of the parish church and on the nearest part of the built up area. As can be seen from the enlarged inset, these are quite clearly 'cementation furnaces'. The implication is that there were only two small units of this type in the town at this date; in view of the later importance as a steel centre this will probably come as a surprise to many people.

3 Sheffield from the Attercliffe road, 1819. It was in this year that the canal was finally brought right into the town; considering the state of the main road from Tinsley as seen here, the improvement in transport facilities was highly desirable. Signs of industrial activity within the town are quite evident. Rather ironically, this stretch of the river bank is still known as Salmon Pastures. It must be many a year since a salmon was seen; nevertheless, while the story of the eighteenth-century apprentice being promised he would not be fed with salmon more than three times a week may well be folklore, there is in existence a lease of the Blonk Tilt, dating to 1759, requiring the annual payment of a dish of fish from the dam to the Duke of Norfolk as landlord.

4 *(opposite)* **Prospect of Sheffield,** 1855. Here we can see the canal (with the basin in the centre of the picture) and the Sheffield, Manchester and Lincolnshire Railway (dating from 1849, later to be incorporated in the Great Central Railway). A train can be seen in the right background, just having left Victoria Station and crossing the viaduct over the canal. Conical cementation furnace chimneys are now very much in evidence, even in the town centre; in the foreground are grinding wheels, having been hewn out of the local gritstone, and we can see one load on its way to the works in the town below.

5 View of Sheffield. This is taken from a painting of about the same date, giving a view from the western outskirts, showing a quarry with grinding wheels being fashioned from the Millstone Grit. Near the centre of the picture is an isolated steelworks and the industrial smoke from the town centre, here seen as a background, can be glimpsed.

6 *(opposite)* **Sheaf Works, about** 1850. This view is really an enlargement of the extreme right-hand portion of Fig 4, but drawn somewhat earlier before the erection of the Victoria Station, the original Sheffield station being at Bridgehouses, about half a mile further west. The viaduct over the canal can be seen quite clearly; in this case the train is going in the opposite direction. Several interesting features arise here; in addition to the cementation-furnace chimneys, there are sailing barges on the canal, an open coal-mine, a horse-drawn wagon on the tramway on the near bank of the canal and the tow-path on the far side. The canal basin is immediately to the left off the picture, the canal turning through almost a right angle at the entrance.

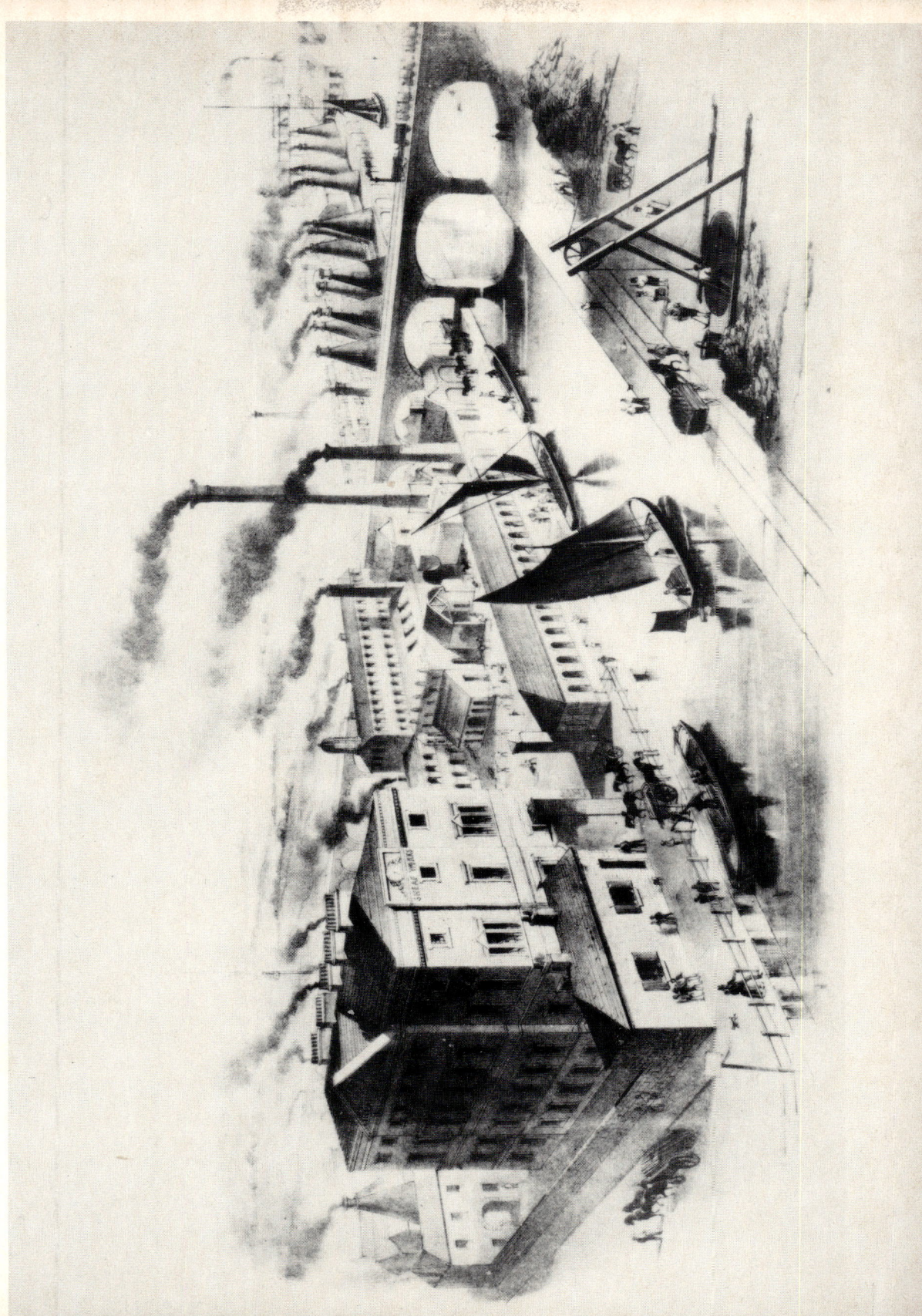

The Cementation Process

7 Converting Furnaces at the Norfolk Works of Thomas Firth and Sons, about 1900. These furnaces were used for 'converting' iron bars into blister steel; they are also known as 'cementation furnaces', since the process by which carbon is introduced into the iron was termed 'cementation'; they should not be confused with 'Bessemer Converters' which were quite different (see Figs 89-91). At one time there were well over two hundred of these conical chimneys in the Sheffield area.

8 *(opposite)* **The Cementation Furnace or Converting Furnace.** This illustration is taken from a report which appeared in France in 1843. The author, Professor le Play, had been commissioned to visit the Sheffield area to examine British steelmaking methods. His report, whose title may be translated as *The Manufacture of Steel in Yorkshire*, is a first hand (and first rate) account and is the earliest detailed source of information on the Sheffield process. We can observe the two sandstone chests, with an arrangement of flues around them such that the flames from the coal fire on the firegrate below could completely surround them and their contents. The conical chimney, in these drawings, stands 40ft above ground level and each chest would be capable of holding about 9 tons of iron bars packed in charcoal.

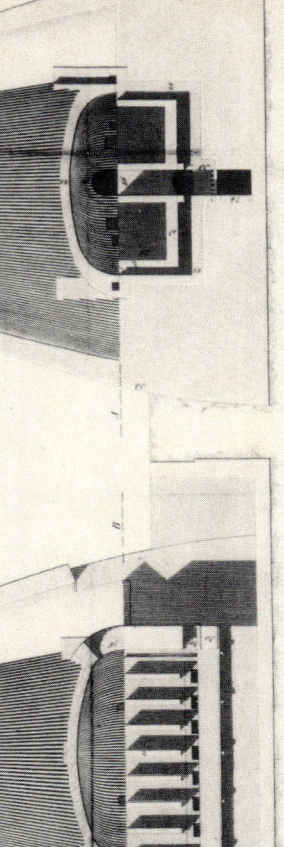
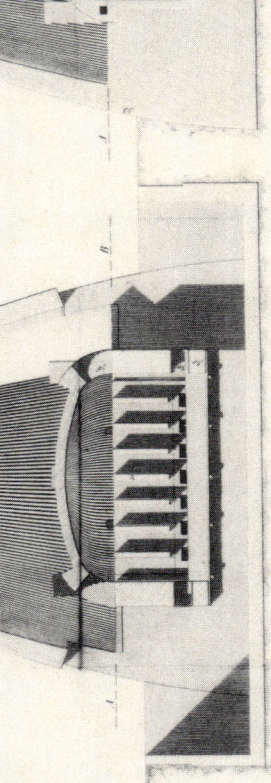
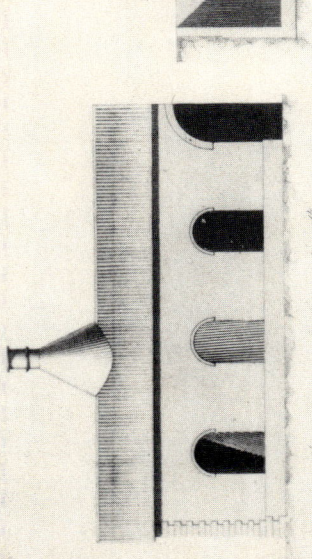
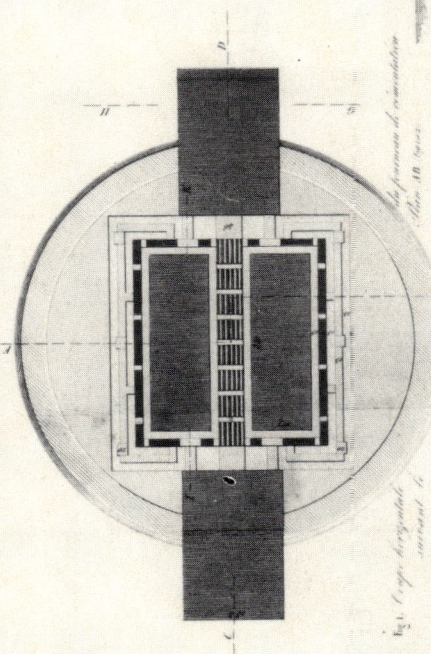

9 Mouth of Converting Furnace, about 1900. Here we are looking into a furnace from the outside; we can observe the main archway, allowing a man to enter the furnace chamber, and the two small holes through which the bars could be passed in to him from the outside, and through which they would be passed back to his mate, some three weeks later, after conversion to steel. Bars ready for charging can be seen stacked outside the furnace.

10 Interior of Converting Furnace. The central flue and the two chests can here be seen quite clearly; note also the side flues with the domed roof over them. The effect of heat on the surface of the brickwork, giving a 'glaze' to it, should be noted. Comparison with the drawings in Fig 8 will enable a clear picture of the internal furnace structure to be obtained.

11 Interior of Converting Furnace. This is a very similar view to the previous one, but shows one chest filled with bars and covered with charcoal, awaiting the final covering with wheelswarf, while the filling of the other chest is not quite complete. Note the method of packing the bars, with small spaces between them so that charcoal can get between and thus surround each bar. The watering can was used to moisten the charcoal and to 'slack' the dust. The cramped working conditions shown here were perfectly typical; the workman is awaiting the passing in to him of further bars by his mate, who can just be seen outside.

12 Early Sheffield Converting Furnace. Taken from a French report of 1774, this shows a furnace with one chest only (holding about 3 tons of iron) at a time when the North-East, producing much more steel than Sheffield, used furnaces with two chests (holding 10 tons between them). It should be remarked that at the same date there were more elaborate furnaces in Birmingham and the Stourbridge area with three chests each and two firegrates, although the total capacity was under 10 tons per furnace.

13 The Doncaster Furnace, Hoyle Street. Out of some 260 furnaces built in the Sheffield area, this is now the only one which is still virtually complete. It was originally part of the Daniel Doncaster works and was probably erected between 1830 and 1840, presumably on the site of a smaller furnace, only half a mile from the town centre. The furnace was damaged in an air raid in 1940 and the top portion was rebuilt in the form now seen, complete with an anti-glare blackout device! Both chests included, the capacity is about 35 tons. The last heat on this furnace, in 1951, was recorded on film.

14 Bower Spring Furnace. This furnace, slightly nearer the centre of the town than the Hoyle Street furnace and less than a quarter of a mile away from it, to the rear of Gibraltar Street, was one of a pair built around 1830 with a total capacity per furnace of 20 to 25 tons. It was originally worked by Thomas Turton and Sons who sold it in 1860 to Moss and Gambles, who seem to have operated it until World War I. It has only recently been excavated; one chest is still in situ and the central flue, together with the back wall flues and chimneys, are quite well-preserved. The Department of the Environment has now spent some effort in consolidating what remains, since it is easier to obtain a picture of the internal construction of such a furnace from these partial remains than from a complete structure.

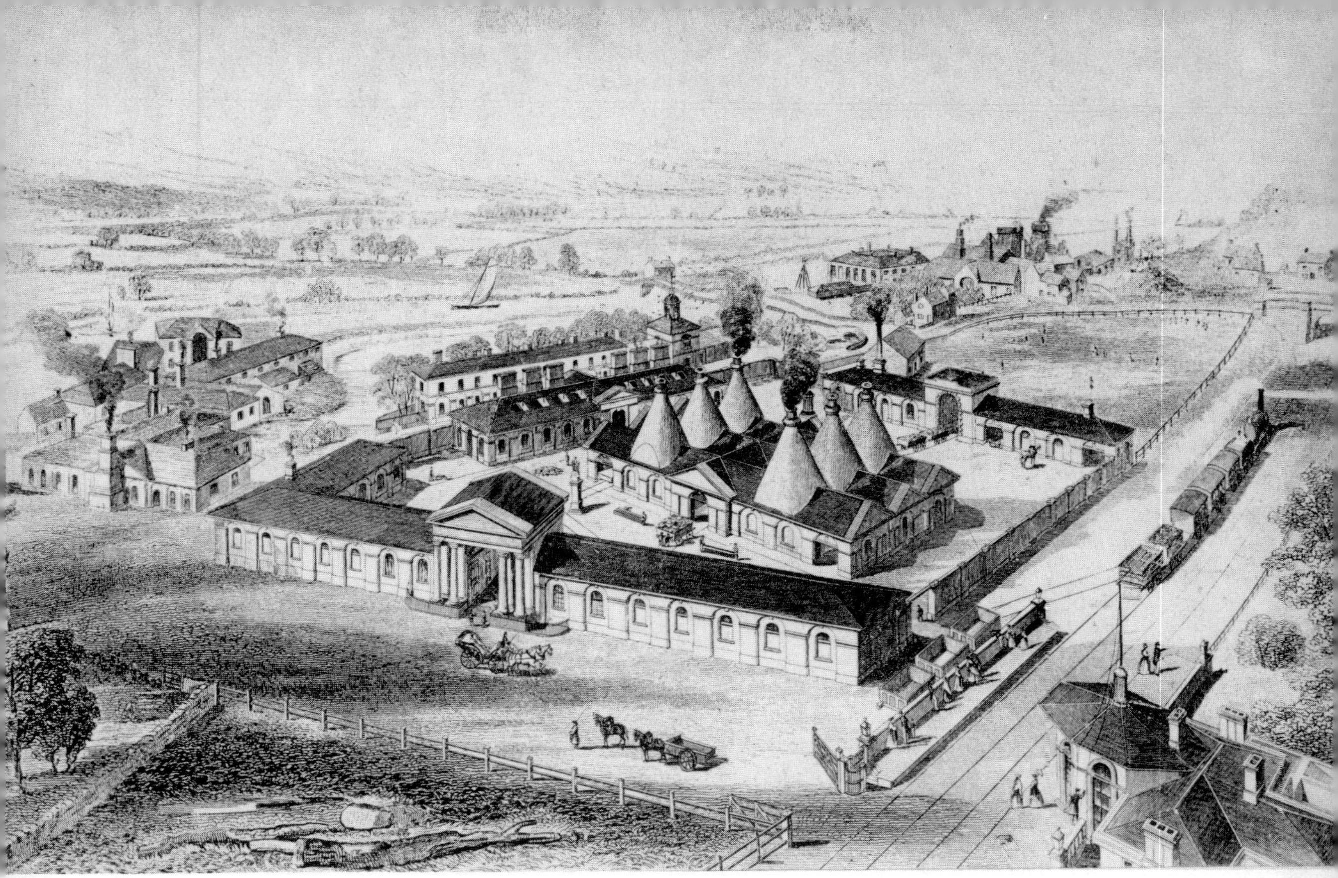

27

15 *(overleaf, top)* **Holmes Works, Rotherham.** This group of six furnaces was still standing until 1969. The site was taken over by Peter Stubs, file maker of Warrington, in 1842, four years after the building of the Sheffield to Rotherham railway. Here we see a train on that line just leaving Holmes station. It is said that Peter Stubs wished to make himself independent of the Sheffield merchants and he toured the area, looking at transport facilities. His final choice gave him the canal, to bring in his raw material supplies of iron from Hull (note the sailing barge in the background) and the railway to take away his product and to provide him with local coal; remember that the Sheffield to Manchester railway was also under construction at this time. Here we have an integrated steelworks, with cementation furnaces to provide the blister steel, crucible melting shops to produce the ingots and forge shops to handle them. This view is of interest also in that it is the architect's impression of the works he was commissioned to erect.

16 *(overleaf, bottom)* **Holmes Works, Rotherham.** The six chimneys can be seen in two rows of three, separated by a central carriageway to service loading, firing and unloading. Each furnace had a total capacity of about 27 tons, so that the works could have dealt with between 2,500 and 3,000 tons of steel a year on full output. Comparison of this photograph, taken a few weeks before the demolition, with the previous illustration shows the former to have been a very accurate one.

17 Holmes Furnaces, Rotherham. This gives an idea of the type of chimney which originally crowned the conical stacks both at Hoyle Street and at Bower Spring. At the Holmes Works the top of the chimney was 56ft above ground level and had a central hole of just under 3ft in diameter. After demolition, it was found that the stones making up the top ring were very finely chiselled on the outside and most carefully dovetailed together, all the more surprising since none of this was ever likely to show from below.

18 Holmes Furnaces, Rotherham. A view looking from the outside of one furnace, down the line of the central flue, with the furnace opening of the corresponding furnace in the other row showing in the distance. The charging openings along the central carriageway can be clearly seen.

19 Derwentcote Furnace, County Durham. This furnace, probably dating from around 1740, although outside the Sheffield area, is worth including as the only virtually complete cementation furnace remaining in approximately its original condition in Britain. It is the last surviving evidence of the extensive cementation steel making in the North-East in the first half of the eighteenth century. It is of the same general type as the Sheffield furnaces, having two chests with their associated flues and firegrate beneath a conical chimney, but this latter was built of stone, not, as in South Yorkshire, of brick. The furnace, which was described in one of the Swedish travel diaries in 1754, is situated near the banks of the River Derwent, a tributary of the Tyne, some twenty miles from Newcastle; it had a capacity of about 10 tons total.

20 Holly Street Furnaces, Sheffield. Here, around 1800, we find two converting furnaces actually attached to a dwelling house; similarly, it is recorded that Huntsman's works were approached through his living room. It must, at any rate, have made security a little easier!

The Supply of Iron

21 *(opposite, top)* **Firth's Iron Wharf.** It must be appreciated that the import of iron from Sweden was an essential part of the old Sheffield steelmaking activities. This iron was made from high-grade iron ore mined in the Dannemora region in Sweden and this was smelted and refined to bar iron using charcoal exclusively as fuel, there being no coal available in Sweden. It was thus a very pure iron, particularly low in its content of those two detrimental elements sulphur and phosphorus. Many thousand tons of iron, in rectangular bar form (a general size would be $3\text{in} \times \frac{5}{8}\text{in}$) came up the canal into Sheffield every year and many firms had their own wharves. This shows the Thomas Firth and Sons' Wharf about 1900.

22 *(opposite, bottom)* **Firth's Iron Store, about 1900.** The more prized varieties of Dannemora or Oregrund bars were kept under cover; here we see a stock of about 2,000 tons in a shed adjacent to the wharf shown in Fig 21.

23 Firth's Iron Yard, about 1900. Out in the yard at the back of the wharf, the 'commoner' Swedish iron grades (although prized much more highly than any local supplies) were treated with rather less care and were left out in the open; there was a reasonably rapid turnaround, but one suspects that some of the 6,000 tons here could have spent a fair period of time exposed to the atmosphere. Being wrought iron, however, it would not rust excessively.

24 *(opposite, top)* **Jessop's Swedish-Iron Store.** Roughly contemporary with the last three illustrations, this shows a very similar arrangement for premier-quality iron. It should be noted that the lengths are all standard – to fit the cementation chests – and likewise the bars are all straight and nicely flat. Note further that the origin of each batch is clearly preserved, presumably in case of any dispute over quality.

25 *(opposite, bottom)* **Sheffield Canal Basin, about 1910.** Apart from the biggest works, a number of firms bought from merchants who had warehouses elsewhere in the town. Some idea of the activity on the canal can be obtained from this view, believed to date from just before World War I. Once more, we see a sailing barge, this time with sails lowered to enable it to pass under the town bridges.

26 *(left and opposite, top)* **Swedish Iron.** The manufacture of iron in Sweden was closely controlled, both on the grounds of quality and of output. The various works not only had their own trademarks, or 'Stampel' (obviously related to our word 'stamp mark'), but the production from each works in any one year was clearly laid down and monitored. Here we have the title page from the 1845 edition of the *Stampel Bok* (which can be translated as *Trade Mark Register*), together with a specimen page giving the trade mark or stamp, the firm, the ironmaking plant available, the permitted output, the owner or agent and other relevant comments. Some of these marks, particularly Ⓛ (Hoop L), from Leufsta Forge, and O-O (Double Bullet), from Osterby Forge, were highly prized for over two centuries.

27 'Hoop L' Iron. This is a photograph of an actual sample of this famous brand, which is still held by Daniel Doncaster and Company. They acted as main importers of Swedish iron, in addition to being steelmakers. Every bar of iron coming to Sheffield from the Leufsta Forge had this stamp on it. So much was it in demand that certain unscrupulous merchants from time to time stamped other iron with this mark. This sort of thing led eventually to major contracts being drawn up directly between the Swedish forges and the Sheffield steelmakers, so as to obviate such fraud.

Bruks- eller Smides-Stämplar	Brukens namn och belägenhet	Härdar	Hammarskatt			Privilegieradt årligt Smide			Utskeppnings- eller Afsättnings-orten	Särskildta förhållanden i anseende till Smide och Verkstäder	Tackjernstag för Smidet	Ägare eller Bruks-Disponenter	Annotationer
			Sk℔	L℔	℔	Sk℔	L℔	℔					
	Upsala Län.												
(C with crown)	**Carlholm** se Löfsta.												
	Elfkarleö i Elfkarleby Socken.	4	18	10	—	1,850	—	—	Stockholm.	Spiksmide finnes äfven vid Bruket.	Eget tackjern	C. Tottie.	
(G with crown) Bistämpel: SYKES	**Gimo** i Skefthammars Socken.	2	3	15	—	1,500	—	—	Stockholm.	Hammarskatts-smidet utgör endast 375 Sk℔, men smidesrätten är oinskränkt, mot årlig Recognitions-afgift af 133 Rdr 16 sk.	Eget tackjern	Disponent: Friherre A. Reuterskøld.	
(L with crown) Bistämpel: SYKES	**Löfsta** och **Carlholm** i Löfsta och Westlands Sr.	6 2	} 20	—	—	6,000	—	—	Stockholm.	Egentliga Hammarskatts-smidet är 2,000 Sk℔, ehuru smidet för öfrigt är obegränsadt, emot årlig Recognis-afgift af 533 Rdr 16 sk.	Eget tackjern	H. Exc. Grefve Carl de Geer.	
(W with crown)	**Strömsberg** och	2	16	10	—	1,650	—	—	} Stockholm.	Spik- och Manufactur-smide finnes äfven härstädes.	Eget tackjern	H. Exc. Grefve Carl de Geer.	
(W with crown)	**Ullforss** i Tolfta och Tierps Socknar.	2	14	10	—	1,450	—	—				Disponent: Grefve B. von Platen.	
(S) (S C G) Stämpel för Ankar-smidet.	**Söderforss** i Söderforss Socken.	3	22	2	13	2,453	6	—	Gefle och Stockholm.	Här finnes 240 Sk℔ så kalladt Frälse-smide, men oinskränkt Ankar-smide vid särskildte Verkstäder, mot Hammarskatt, samt diverse Manufactur-smide.	Eget tackjern	Interessenter. Disponent: Friherre P. A. Tamm.	
	Ullforss se Strömsberg.												
(B)	**Wattholma** i Lenna Socken.	2	7	—	—	1,400	—	—	Stockholm.	Det här varande högre Frälse-smidet utgör 700 Sk℔.	Eget tackjern	Disponent: H. Exc. Grefve M. Brahe.	
(P with crown)	**Åkerby** i Löfsta Socken.	2	20	—	—	2,000	—	—	Stockholm.	Stångjerns-smidet är, mot skattefrihet, inställdt på bestämd tid; men Stål-tillverkning drifves på Åmesjern från Löfsta.	Eget tackjern	H. Exc. Grefve Carl de Geer.	
(OO) Bistämpel: SYKES	**Österby** i Films Socken.	4	15	—	—	2,600	—	—	Stockholm.	Hammarskatts-smidet 1,500 Sk℔, emot årlig Recognis-afgift af 533 Rdr 16 sk., är smidesrätten oinskränkt; och drifves dessutom en betydlig Stål-tillverkning.	Eget tackjern	Friherre P. A. Tamm.	

*) Medel-tillverkningen är här uppgifven, emedan smidet vid tillåtne Verkstäder är fritt.

*) Frälse-smide, eller Frälse-rätt kallas det skattefria smide, som, enligt Kongl. Brefvet d. 2 Aug. 1731, [...] vid [...] i allmänhet gagnas till ett belopp af 15 procent; men vid några Bruk finnes dock ett så kalladt högre Frälse[...]

28 Other Dannemora Irons. These illustrations are taken from a paper on Sheffield steelmaking published in 1869. 'Steinbuck' was included as representative of the 'commoner' grades of Dannemora iron, selling at around £24 per ton; 'W and Crown' (here marked as Bar Iron) was a 'middle' grade, worth about £29 per ton, while 'GL', 'Double Bullet' and 'Hoop L' would fetch up to £34 a ton. The steelmaker, in pursuit of quality, paid such prices even when he could get English bar iron (for example, 'Best Yorkshire') at £18 per ton or less.

Steinbuck.

Hoop L.

GL.

Bar Iron.

Crucible Steel Melting

29 Benjamin Huntsman's House at Handsworth. The inventor of the crucible process is well known by name; being a Quaker, however, he refused to have his likeness portrayed, so we shall have to be content with an old picture of his home at Handsworth, then a village four miles or so from Sheffield. He moved here in the early 1740s from Doncaster. These premises were demolished early this century, but it was reported that flue marks could be traced in the small building at the end of the house, so this may have been the place where it all started!

30 Ersta Steelworks, Stockholm, 1769. We have no record of Huntsman's works either at Handsworth or at Attercliffe, where he moved in 1751. We do know, however, from the report of Robsahm, a Swedish visitor to Attercliffe in 1761, that he was melting about 13lb of steel in each crucible. Six years later another Swede, Bengt Quist Anderson, visited the same premises; two years later he set up a crucible-steel melting shop near Stockholm and here we reproduce a plan of those works, with two rows of three furnaces each, the chimneys being about 30ft high. Huntsman's own shop cannot have been very dissimilar to this.

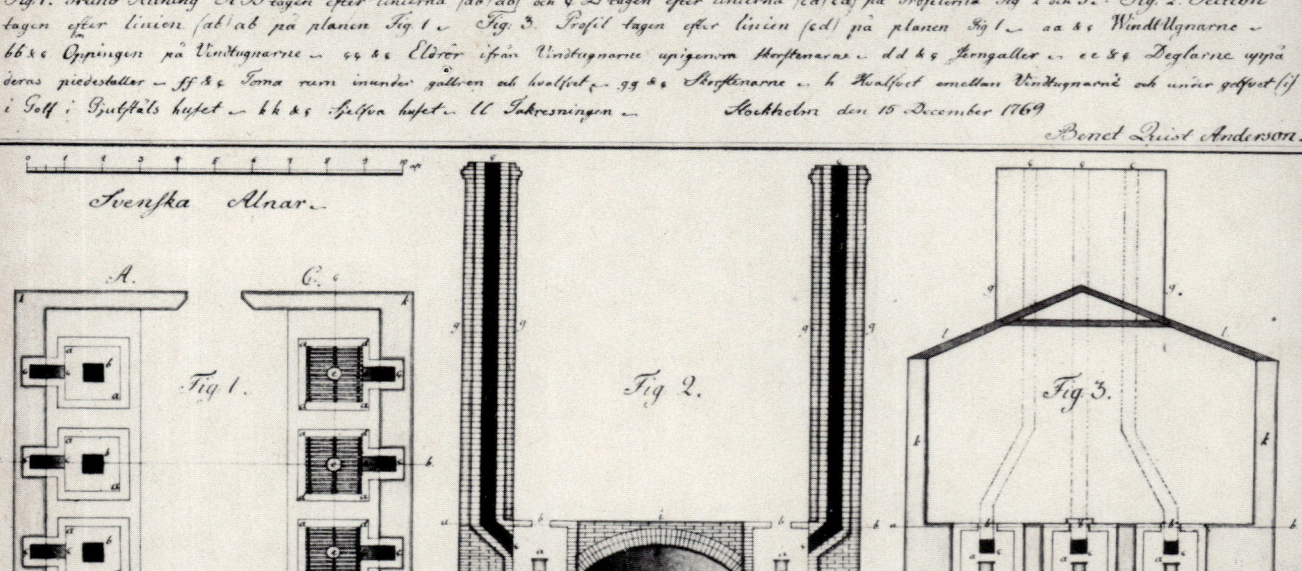

31 A Steelworks near Sheffield, 1797. Some thirty years later another Swedish visitor came to the Sheffield area and his report contains this most intriguing drawing of a crucible steel melting shop 'near Sheffield'. From the text it seems likely that it was the Walkers' premises near Rotherham, but it could just be the Huntsman establishment at Attercliffe, since this was also 'near Sheffield'; these seem to have been the only two works of this type which he visited. At any rate, all the necessary operations were going on in a peaceful rural setting. The large cellar, to provide the necessary air supply for the combustion of the coke, and the rectangular chimneys, each providing the draught for five melting holes, are clearly seen. The man in the cellar is preparing the charges; upstairs we can see a crucible being drawn from a furnace and an ingot being cast. The same report gives details of the ingot-moulds and the tongs used; it would appear that the charge weight was around 25-30 lb, roughly double the scale of operations thirty years earlier.

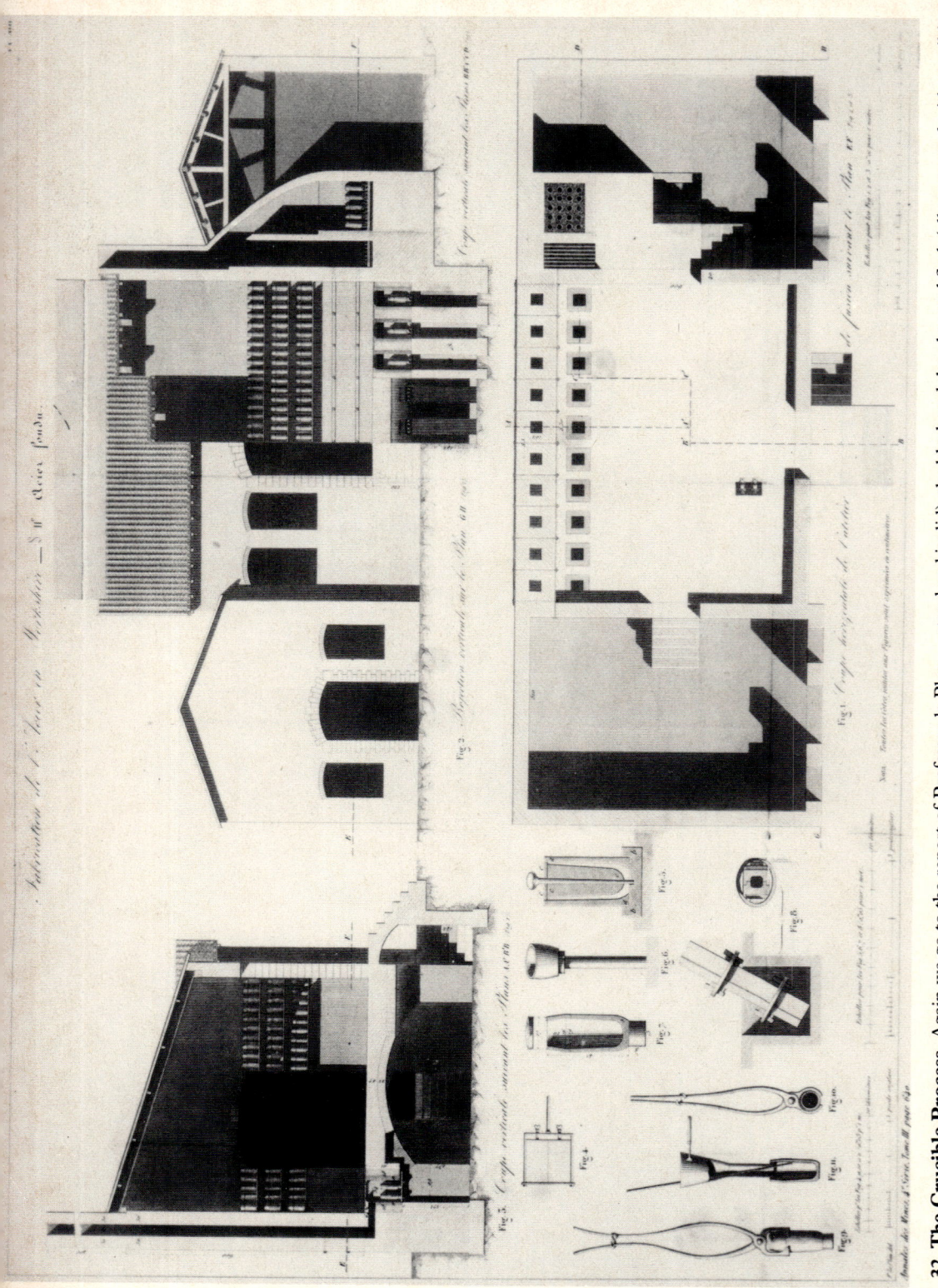

32 The Crucible Process. Again we go to the report of Professor le Play, published in 1843, for this illustration. The buildings themselves, which are stated to be typical of the best layouts in Sheffield, depict a ten-hole furnace shop. We can also observe in the group in the bottom left-hand corner, on the top row, reading left to right, the furnace-hole cover (an iron frame, fitted with firebricks and provided with a handle), a complete crucible (with its pedestal or stand and its lid), the block and the plug and flask (all connected with crucible making, of which more later); on the bottom row in this group we have the lifting tongs for drawing the crucible from the furnace, the charging funnel, the tongs used for teeming the metal into the ingot mould, the ingot mould itself set up to receive the metal and a top view of the mould with its clamping ring and wedges.

33 Crucibles in the furnace. This is not from Sheffield, but is included since it may be considered as typical of what might have been observed in the newer crucible furnaces installed after about 1870. It comes from Pittsburgh and the photograph was taken as late as 1940, just before operations ceased and shows the view into a gas-fired furnace, as distinct from the traditional coke-fired ones. Eventually there was a considerable number of such furnaces in the Sheffield area; they were cheaper to run, but there was considerable prejudice against them from the old hands in the business.

34 Crucible making: mixing the clay. Most works made their own crucibles, since an unfired clay crucible would not travel and a fired one was useless if left to go cold. Here at the Brightside Works of William Jessop in 1900 it was stated that they used 1,000 'pots' (crucibles) a week. Since a crucible would be used for three melts only, it can be calculated that the annual output was of the order of 3,500 tons, which fits the production of their cementation furnaces. It should be made clear, however, that their crucible steelmaking capacity forty years earlier was at least 50 per cent greater. The material for the crucibles was a carefully weighed (or measured) mixture of local Stannington clay with imported Stourbridge clay, a little China clay, together with some fired clay, carefully ground up and known as 'grog', and a proportion of coke breeze. Tradition has it that Huntsman himself incorporated these last two ingredients to prevent the collapse of his pots on firing. The dry ingredients were put into the treading tray, mixed well, treated with a measured volume of water and left overnight for the water to be absorbed.

35 Crucible making: treading the clay. Next morning the clay would be well mixed and turned over with shovels and it then had to be trodden, to mix it even more thoroughly and also to drive out air bubbles. It was also said that the bare feet could discover sharp angular pieces in the clay which could have caused failure of a crucible and which could be picked out at this stage. One wag even said that to use bare feet was the handiest way of doing the job! In smaller works, such as the Abbeydale Hamlet, only one man would have been involved; here, at the Jessop works, there are four men.

36 Crucible making: the clay ready for moulding. Again in the Jessop shop we can see the batch of trodden clay. Note the crucible-making machine to the left of the bench behind the treading tray and a number of pots on the shelves, drying out slowly. These Jessop illustrations all date from the first few years of this century.

37 Crucible making: balling the clay. The correct amount of the prepared clay was weighed out, rolled out with the hands on the bench top, knocked back on its ends and eventually shaped into a rectangular block with a slightly pointed end, as can be seen here in the Jessop works. Note the wooden boards under the bench which will eventually support individual freshly-moulded crucibles until they have dried off sufficiently to handle.

38 Crucible making: moulding the crucible. The workman in the centre is placing the 'plug' into the 'flask' which already contains one of the balled lumps of clay. The outer surface of the plug and the inner surface of the flask were brushed with oil before use to facilitate a clean strip of the moulded crucible. The plug was then hammered home with a mallet and just a small amount of clay would ooze out round the rim. This was cut off with a knife and the plug lifted out. The flask was then lifted up and placed on a cast-iron block set up in the floor and the formed crucible, standing on the loose base plate from the flask, would be revealed as the flask slid down to the floor. The crucible was then carefully picked up between two suitably shaped pieces of sheet iron and placed on the wooden stand on the bench. This is another of the series of Jessop illustrations. Fig 32 gives further information on the equipment.

39 Crucible making: finishing the crucible. The plug and flask produced a wide-mouthed crucible; here the 'bonnet', made from thin sheet-iron, is being used to bring the top to a reverse taper. The bonnet was slowly rotated with a slight downward pressure. This is a late photograph, dating from about 1960, showing the last crucible-maker at the Huntsman works in operation.

40 Crucible making: machine moulding. In the Jessop works, as in other large organisations, it was eventually found imperative to speed up the crucible production and machines of this type were installed, working on precisely the same principle as the hand-operated method, but driving in the plug with a screw press.

41 *(opposite, top)* **Weighing the charge.** The material to be melted was weighed out carefully in pieces of a size suitable to charge in the crucible. In the early days, up to about 1850, the charge would almost certainly have been blister steel, of the appropriate carbon content as judged by its fracture. Gradually other mixtures would be used, still with material of Swedish origin. Rather than putting the iron through the cementation furnace, however, mixtures of the iron with charcoal or, better still, mixtures of the bar iron with Swedish white cast-iron (containing about 4 per cent carbon, but still low in the deleterious elements sulphur and phosphorus) were charged to the crucibles. This illustration again comes from the Jessop works.

42 *(opposite, bottom)* **'Recipes'.** This reproduces a circular issued by Daniel Doncaster and Sons about 1880. This firm not only produced crucible steel but imported Swedish iron for resale. The material here quoted as 'Warner's Carbonising Iron' was a Swedish white cast-iron of the type mentioned above; 'Spiegel Eisen' was a similar material but containing in addition about 10 per cent of manganese, generally considered to be a beneficial addition to steel in small quantities. Note how the carbon content of the melt is adjusted by altering the proportions of the two irons.

RECIPES

FOR

THREE MAIN OR STANDARD TEMPERS OF STEEL,

From which any intermediate tempers can afterwards be arrived at.

No. 1 FOR SAW FILES, &c.,

CONTAINING 1.50 % CARBON.

24 lbs. Swedish Bar Iron.
20 „ Warner's Carbonizing Iron.
1 „ Speigel eisen.

No. 2 for TOOL STEEL for TURNING, &c.,

CONTAINING 1.16 % OF CARBON.

34 lbs. Swedish Bar Iron.
15 „ Warner's Carbonizing Iron.
1 „ Speigel eisen.

No. 3 FOR SAWS, TABLE KNIVES, &c.,

CONTAINING 0.75 % OF CARBON.

40 lbs. Swedish Bar Iron.
9 „ Warner's Carbonizing Iron.
1 „ Speigel eisen.

The Speigel is added for the purpose of neutralizing the sulphur, every part of which requires 7 parts of Manganese to neutralize it.

43 Charging the crucible. The crucible was brought up to a red heat the night before it was required. Next morning it was placed on its stand on the firebars of the furnace, its lid put on, a few live coals placed around it, followed by a charge of coke to fill in the rest of the space, the furnace cover placed on and the full draught applied so as to heat it to a full working temperature. The furnace cover was then taken off, the crucible lid removed and the charging funnel placed into the mouth of the white-hot crucible. Generally a long iron bar was first put through the funnel to the bottom of the crucible, to break the fall of the metal charge which was then gently slid into the funnel from the charging pan. The bar and the funnel having been removed, the crucible lid was replaced.

44 Charging up with coke. The gap all around the crucible was then filled up with coke, the furnace cover replaced and the draught re-applied. Two further charges of coke would be required, roughly at hourly intervals, before the metal was fully melted, and a further charge would then be used to 'kill with fire' before the metal was ready for withdrawal and pouring into the ingot mould. These last two illustrations were taken about 1960 in a small melting shop near the centre of the town, just before these operations finally ceased.

45 Huntsman's crucible shop. The general scale of operations in a medium-sized shop can be judged from this illustration, taken from the Huntsman catalogue of 1930, but dating back many years, probably representing the shop in pre-war days. Note two charge-pans lined up in front of each furnace to satisfy the two pots in each hole on recharging. Note also the rows of pots drying on the shelves over the furnaces. On the right, the metal in the crucibles is being 'tested' with an iron rod; in this way it could be assessed whether all the charge was melted and, if so, the time for the end of the rod to melt off gave an indication of the temperature of the metal. An ingot mould is being set up on the left of the picture.

46 Marsh Brothers, Pond Steelworks, 1897. This was another shop on a similar scale. This view shows the method of setting up the moulds quite clearly. Note the traditional style of wicker-work coke basket (one full and one upside down) with a furnace cover in front of them. In the back ground there are two crucibles which have just been drawn from a furnace; the heat from the open furnace-hole can be sensed.

47 Samuel Osborn and Company. Here we can see a large crucible shop in a photograph of around 1910. There is a long row of crucible holes, with the crucibles drying on shelves above them, a pot just withdrawn, an ingot being poured, metal being tested further down the shop, a coke heap with baskets, and so on, repeating on a larger scale all that has been seen elsewhere. All that could be done with crucible steel to increase the output was to multiply the number of units.

48 Casting the ingot. This delightful print, a very reasonable representation of the casting operation, was drawn in 1941 and used by a local refractories manufacturer on a calendar. It well portrays the essential 'cottage industry' phase of the early crucible steel-making in the town and shows the teeming of the metal from the crucible to the ingot mould. Note the way the teemer uses his thigh as a fulcrum and the weight of his body to balance the weight of metal, crucible and tongs, which together could be up to one hundredweight.

49 Casting the ingot. This is a very late photograph taken in the Firth Brown works when some crucible holes were opened up again during World War II, after ten years or so of idleness, to make the sorely-needed tool-steel. The man on the left is holding a 'dozzle' in his tongs. The dozzle, or hot top, heated to a high temperature, is made from refractory material and fits into the top of the ingot mould, having a central hole about two inches in diameter. When almost all the metal from the crucible has been poured, the dozzle is inserted and the central hole filled up with the last metal. This acts as a reservoir to feed liquid metal down into the centre of the ingot as the metal contracts on solidification, thus giving sound metal below the dozzle. Three fully-cast ingots and an empty mould can also be seen. The photograph was taken without any extra illumination other than the glare from the molten metal, and the atmosphere captured is remarkable in its likeness to the industrial paintings of Joseph Wright of Derby from the 1770s.

50 Casting the ingot. This is another illustration from the Huntsman catalogue. In the centre it would appear that 'doubling-up' is taking place – the pouring of the contents of one crucible into another before teeming, so as to enable a larger ingot, of about 100lb in weight, to be made. This could well have been a rectangular ingot (say 8in × 3in section) from which sheet could be produced, for the manufacture of saw blades, for instance. In the left foreground is the crucible 'barrow', used to transport the freshly-withdrawn crucible across the shop to the casting pit.

51 Casting larger crucible-steel ingots. Here in the Jessop works a large number of crucibles have been made ready at one time. The method was to set an ingot mould below floor level and arrange what was known as a 'tundish' – a deep dish-shaped iron vessel with a refractory lining and a nozzle set in the centre – immediately above the mould. The contents of a number of crucibles could then be poured into the tundish in such a manner that it would never drain completely until the mould was full. In this particular case it appears that an ingot of two or three tons in weight was being made, requiring the contents of up to a hundred crucibles.

52 Casting large crucible-steel ingots. This famous drawing depicts the casting of an ingot of about 20 tons in weight by the same principle. The largest recorded ingot of this type required the pouring of the contents of no less than 672 crucibles to provide a 25 ton ingot at the huge River Don Works of Naylor, Vickers and Company. The details relevant to the particular operation shown here were not recorded – it was drawn on the occasion of the visit of the Prince of Wales to Sheffield in 1875 – but we have a report of the casting of a gun ingot from 584 crucibles in the same Thomas Firth Works a year earlier. The atmosphere created by the artist is most realistic; the whole concept required the skilled co-ordination of a large army of men and any false move could have brought about injury or death, or could have been detrimental to the production of a sound ingot. It is a pity, however, that he drew the crucibles with round bases – they would never have sat upright in the furnaces!

53 Crucible-steel foundry. As well as using tundish-pouring from several crucibles to produce ingots, the same technique was used to fill sand moulds to produce steel castings. The pioneers in this field in this country were Naylor, Vickers and Company, from about 1854. Many other firms followed suit, Hadfields being famous for their engineering castings. This picture shows the Jessop Foundry around 1900. The mould has been set below the grating in the floor and crucibles of molten metal are being wheeled in from the melting shop as required.

54 and 55 *(opposite)* **The San Francisco Bell.** Naylor, Vickers and Company made a speciality of casting steel bells. These had a different and, it was said, a more pleasing sonority, than the conventional bronze ones and appear to have been cheaper. In 1860, in their River Don works – but at Millsands, not down Brightside Lane – they cast a bell weighing about 5 tons, destined for the Fire Station in San Francisco. Note the two-sided approach to the tundish mouth; it is intriguing to try to count the number of crucibles in the two processions from the tong handles! An attempt to trace this bell recently led to the conclusion that it must have been lost in the earthquake of 1906.

56 and 57 Steel foundry. While strictly speaking this belongs to a later age, since the steel was supplied in bulk by a different process, these illustrations form an interesting contrast. They come from the Jessop Foundry early in the twentieth century. The mould in the foreground was made for the casting of a large spur wheel, the finished steel casting being shown on the left.

Sheffield Steelworks

58 Sanderson's West Street Works. The early steelworks were set up on the edge of the town, very close to what we now regard as the city centre. Here, in West Street, the Sandersons had their establishment early in the nineteenth century. Michael Faraday requested Charles Sanderson to produce experimental alloy steels and they were melted, cast and forged here in 1822. Georg Fischer, the first man to produce crucible steel on the Continent, made many journeys to this country from his native Switzerland and he knew how to find these works when he visited Sheffield in 1826. The bringing together of cementation furnaces and crucible furnaces makes the normal pattern for this time. These buildings were reported still to be standing in 1935, but seem to have gone soon afterwards.

59 Sanderson, Naylor and Company, Millsands. This comes from a print of about 1830. Again we have the combination of cementation and crucible furnaces, together with a forge building, arranged around a central courtyard. Here we find the introduction of another feature which was to be repeated several times elsewhere – an imposing archway entrance with the administrative block flanking it. The location out in the country (although the town was soon to grow beyond it) and the rectangular 'fortress' appearance bring to mind the Roman farm-villa of 1,500 years earlier.

60 Ponds Works about 1846. Here was another old works near the town centre. While most of these old drawings can be taken more-or-less at their face value, it must be admitted that one or two of the 'business card' illustrations, of which this is one, are a little fanciful and the placing of the cementation furnaces does seem to be artistic licence. As far as can be made out from the existing records, and by a comparison of this drawing with later ones, it does, in fact, seem likely that the furnaces shown here did not exist on this site but are the Navigation Works furnaces, which were leased by Marsh Brothers, who ran Ponds Works, until 1852 and were subsequently taken over by Cocker Brothers (see Fig 83).

61 *(opposite, top)* **Killamarsh Steel Works in 1848.** Although not actually in the town, this work is of interest in that it was taken on lease by a steel-user to obtain his own supplies rather than have to rely on merchants – a similar case to that of Peter Stubs and the Holmes Works (see Fig 15). Joseph Webster, of Webster and Horsfall, wire makers of Penns Forge near Sutton Coldfield, held these works from 1824 to 1887. Until 1831, the crucible-steel ingots were forged to billet and then shipped to the Midlands; subsequently, rolling mills were installed at Killamarsh which not only supplied all the requirements at Penns, but developed a market locally for its surplus. The isolated tall chimney was presumably connected with the running of a steam engine to power the mills.

62 *(opposite, bottom)* **The Atlas Works in 1848.** This was the earlier works set up by John Brown in Furnival Street, a small unit in the old town style. When he moved to the much larger site in Savile Street in 1857 he transferred the name. Here again we seem to have a chimney for a boiler to feed a steam engine for some ancillary operations within the works, in addition to the cementation and crucible-furnace chimneys and smaller chimneys which could well have been from forge furnaces.

The Atlas Works,
Furnival Street, Sheffield

63 Huntsman's Works in Attercliffe, about 1850. This is a very nice drawing, showing a variation on the usual theme.

64 *(opposite)* **Norfolk Works, Savile Street.** This is typical of the advertisements put out in the later nineteenth century, giving an interesting view of the works as well as describing the products, embellished with decorative typography. It also, however, again indicates that some caution must be used in assessing such evidence. In the first place, 1840 was the date at which Thomas Firth and his sons set up business on their own at a small premises near the town centre. This particular page comes from a publication issued in 1862, the Norfolk Works having been built on a green-field site down Savile Street between 1851 and 1855. In addition, at the time of this publication, there were firms established either side of the Norfolk Works – Spear and Jackson's Aetna Works on the left and John Brown's Atlas Works on the right – and neither is here indicated!

65 Norfolk Works, Savile Street, about 1870. Another view of the Thomas Firth and Sons works on Savile Street – and another which has to be treated with caution. It is, in fact, two composite halves, which should have the John Brown Atlas Works block between them. The Norfolk Works, founded in 1851, constitutes the left half and is the same as the block shown in Fig 64. Note that three cementation furnaces seem to have gone in the interim and the shape seems to have altered somewhat as far as the remainder are concerned. The right-hand half shows the Gun Works, erected in 1863; the roadway at the extreme right leads to 'Brown's Bridge' over the railway (see Fig 67).

WHITTINGTON WORKS.
1858 – 1906.

66 Whittington Works, about 1870. In addition to their Sheffield establishment, Thomas Firth and Sons held this Whittington Works, near Chesterfield, from 1858 to 1887. The main point to note is the presence of numerous puddling furnaces, recognisable by their tall chimneys fitted with dampers, all belching black smoke. These produced 'puddled steel' as well as wrought iron. Note the importance of railway communication to such a site, as well as the numerous internal rail-links. We also have here an early example of the use of corrugated-iron sheeting as a roofing material.

67 Atlas Works, Savile Street. This is the large establishment set up by John Brown on Savile Street on moving from his other Atlas Works in Furnival Street (see Fig 62) in 1857. He subsequently extended his operations to the north side of the railway, along Carlisle Street, widening the bridge over the railway (henceforth known as 'Brown's Bridge') in 1863 to improve his communications. This drawing dates from 1865 and shows the usual groups of cementation and crucible furnaces but here again, even more so than at Firth's Whittington Works, can be seen the chimneys of puddling furnaces, making wrought iron for plate production as well as producing quantities of puddled steel for less onerous applications where wrought iron was too soft. Note incidentally that the two parts of the Thomas Firth works, depicted as adjacent in Fig 65, are here shown in lighter print flanking the Savile Street block of the Atlas Works; there are again only three cementation furnaces in the left-hand block and they do look rather more squat than the Atlas Works furnaces.

68 Aetna Works, Savile Street, in 1862. Although the caption to the original reads Etna we use the traditional spelling for this famous works. This was the Spear and Jackson establishment, built about 1852 on Savile Street, eventually to be sandwiched between the extensions to the Cyclops Works and the Norfolk Works of Thomas Firth. Neither of these works is shown in the drawing, although 'Cammell's Bridge', built by Charles Cammell in 1845 over the Sheffield to Rotherham railway, is! The chimneys with dampers in the left centre are intriguing. Did Spear and Jackson also carry on puddling on a small scale, or were these connected with forge equipment?

69 Cyclops Works in 1845. This was the first works to be erected on the green-field site along the Sheffield to Rotherham railway, east of the Wicker Station. This view is from Carlisle Street side of the railway, looking south, the buildings shown being along Savile Street. This therefore shows the other side of Cammell's Bridge from that appearing in the last illustration. The main gateways appear to be surmounted with rather elaborate coats of arms.

70 Cyclops Steel Works in 1858. Here we are looking south down Cammell's Bridge from the Carlisle Street end. We still have a rural scene in the foreground, but the buildings on the other side look a little less 'cyclopean', if that term may be forgiven in the present context! Note several other cementation furnaces shown faintly in the background, and also that the Cyclops Works had now spread either side of the bridge. The firm also had depots throughout Europe and in Canada and the USA at this time.

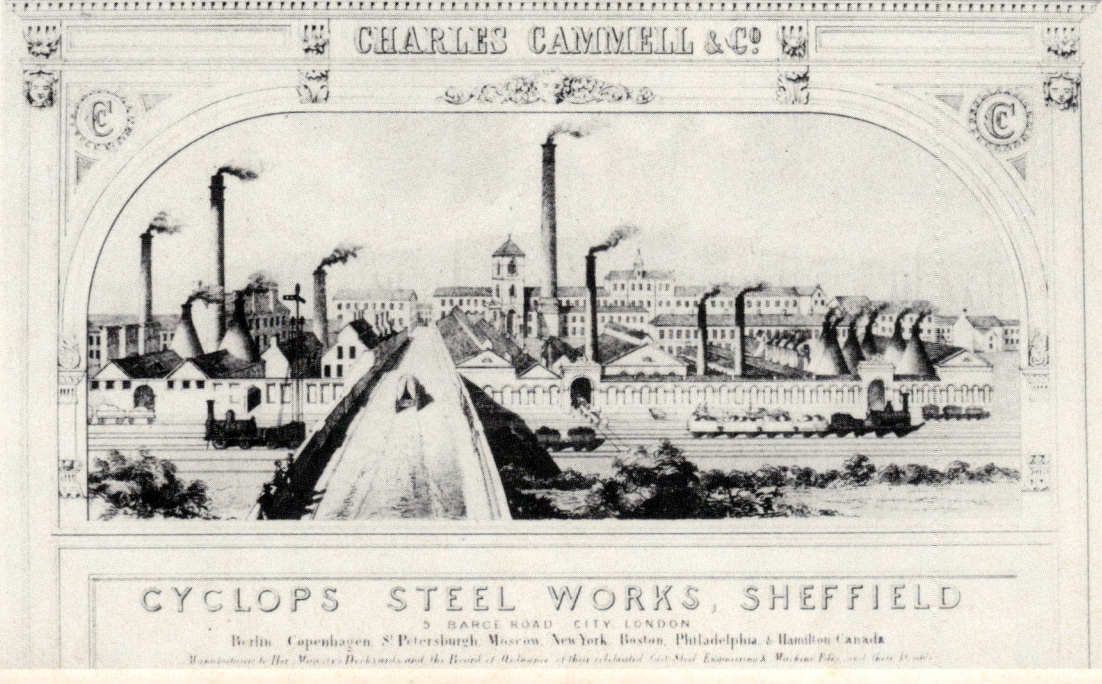

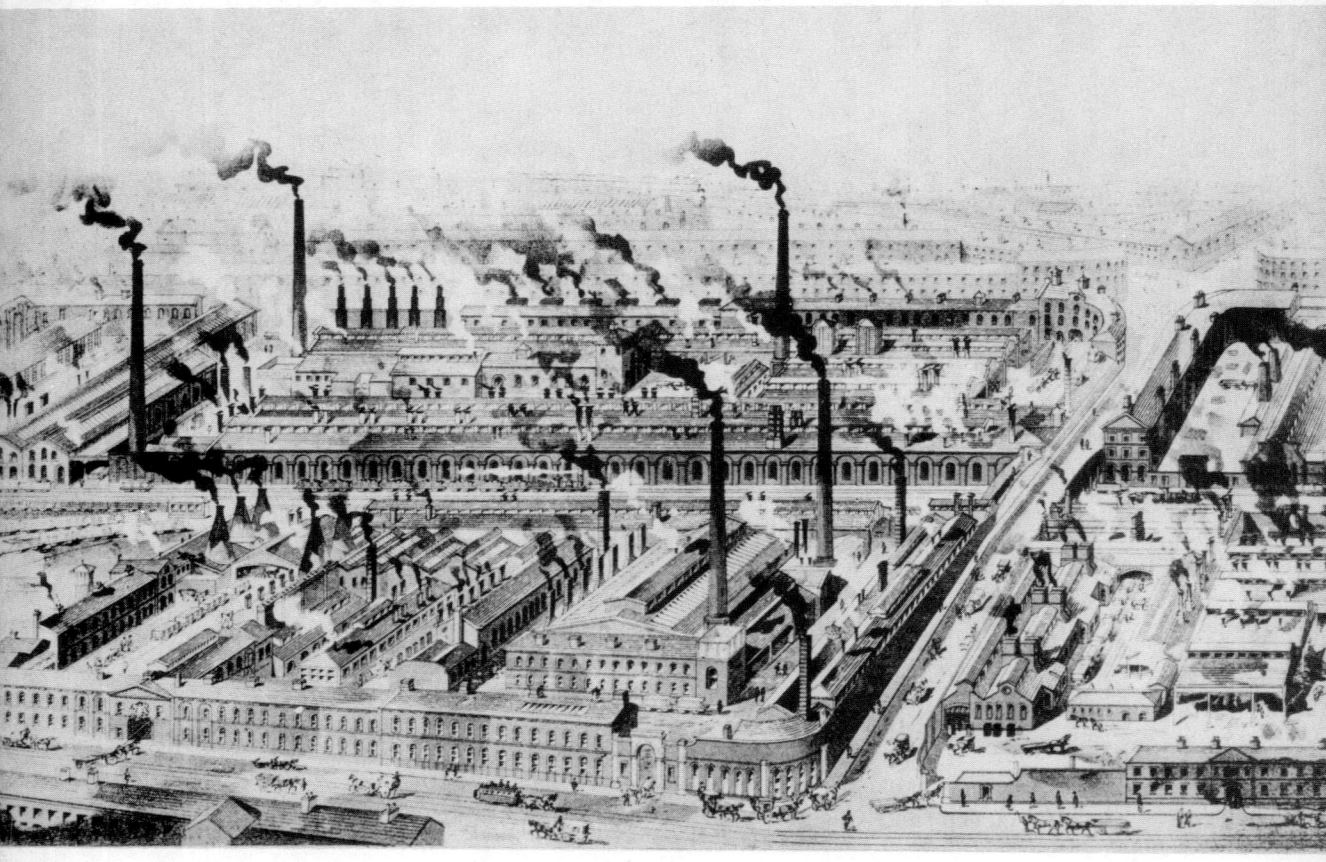

71 Cyclops Iron and Steel Works in 1895. Unfortunately this view is taken looking north; it shows quite clearly, however, that activity has now spread to the northern side of the railway. There have been casualties in the cementation furnace establishment: only five remain while in an 1862 view (not reproduced here) there were thirteen. While there is still evidence of numerous crucible-melting shops, other steelmaking methods were producing the bulk of the product. There is still evidence of puddling furnaces at this date, and it is significant that the title of the works changed over the years (the titles for the last three illustrations have been taken verbatim from the originals) to include iron in the last one.

72 *(opposite, top)* **Sheaf Works in 1855.** We have already had one view of this works (Fig 6) which, if the artist has not misled us, dates from between 1849 and 1851, since there is a railway but no Victoria Station. Here the station can be seen, as it should, since this illustration comes from a letterhead of 1855. The works was built near the canal basin, shortly after the construction of the canal and probably in the early 1820s by William Greaves. It later passed to Eyre, Ward and Company, in whose possession it was at this time.

73 *(opposite, bottom)* **Sheaf Works in 1858.** The works had, by this time, passed to Thomas Turton and Sons who were still also in occupation of the Bower Spring Works (see Fig 14). The collection of cementation furnaces shown here must surely have been the largest in the town. They receive specific mention in the official report of the Paris Exhibition of 1867 as being particularly noteworthy even in the steel town of Sheffield.

SHEAF & SPRING WORKS • SHEFFIELD

THOMAS TURTON & SONS •
MANUFACTURERS OF
STEEL FILES SAWS EDGE TOOLS LOCOMOTIVE ENGINE & RAILWAY CARRIAGE SPRINGS &c
SHEAF WORKS SPRING WORKS AND SLACK STEEL WORKS
SHEFFIELD
AND AT No 17 KING WILLIAM St CITY
LONDON

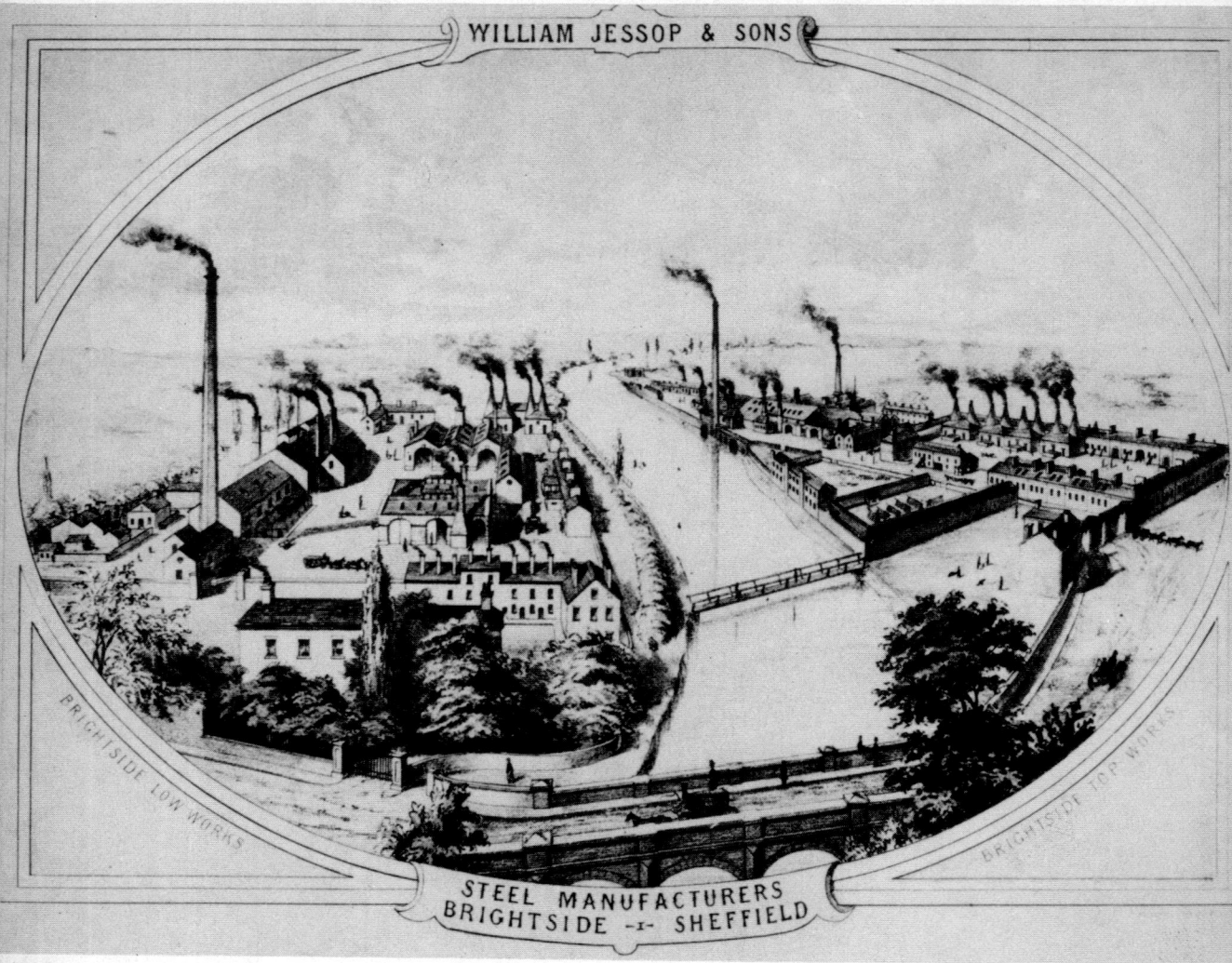

74 William Jessop and Sons, Brightside Steel Works, in 1858. Here we see the Brightside Top Works and Low Works, separated by the River Don but connected by a bridge. The cementation furnaces number four in the Low Works (to the left) with eight in the Top Works (right of the river). The origins of the firm go back to a small establishment near the town centre, first worked in 1774, the location still being known as 'Jessop Street', although now part of a new housing estate. The first activity on the Brightside site was in 1835.

75 *(opposite, top)* **Brightside Steel Works in 1862.** A view similar to the last one, showing the addition of two further cementation furnaces in the Low Works and a building which looks very much like a chapel at the nearer corner of Top Works: could this have been a Church of England mission?

76 *(opposite, bottom)* **Brightside Steel Works in 1879.** The 'chapel' has now gone to make room for rail access to the works. The Low Works now had a range of eight cementation furnaces to match the Top Works. The additional furnaces must have been amongst the last of this type to have been built in the town.

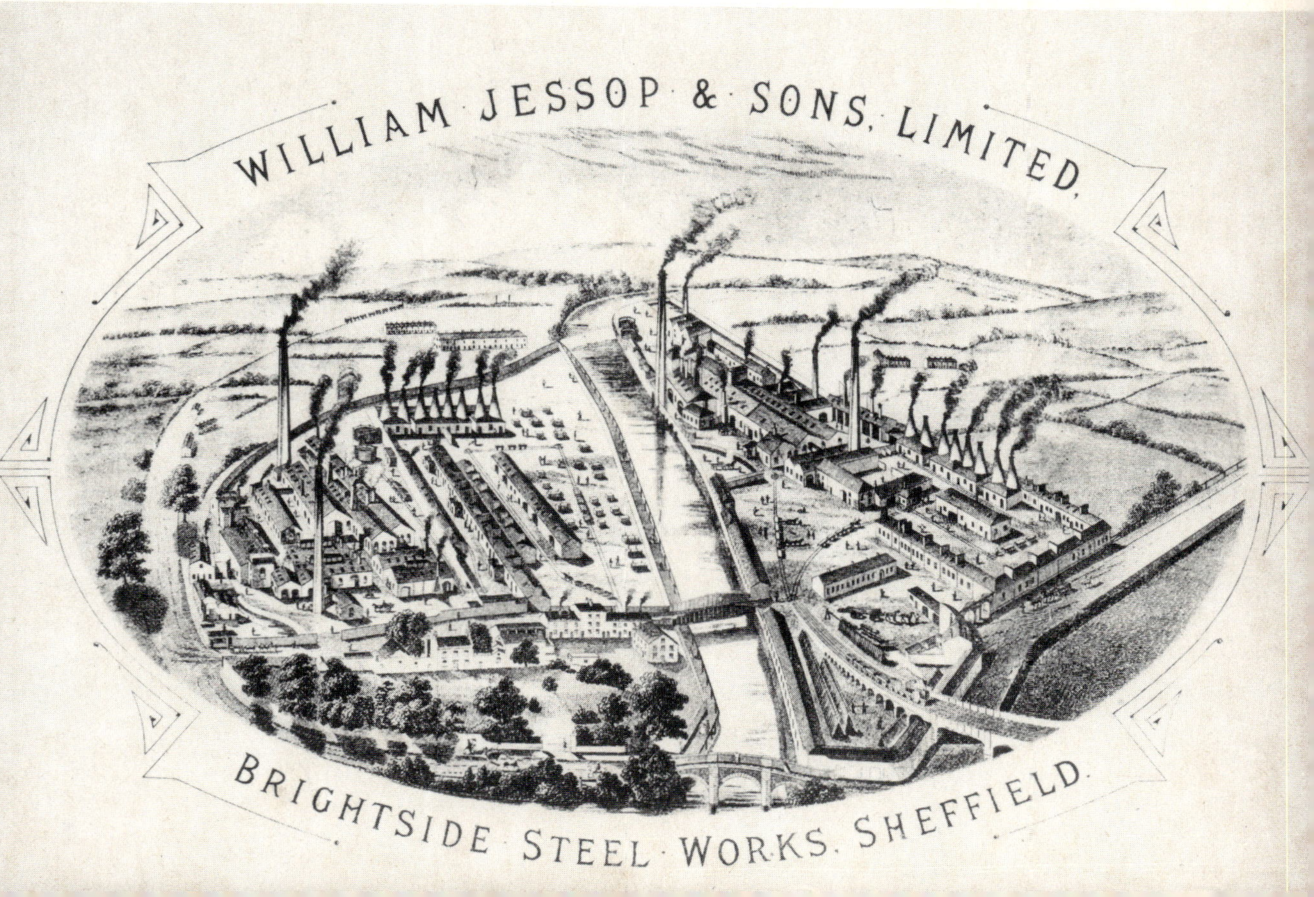

WILLIAM JESSOP & SONS, LIMITED,

BRIGHTSIDE STEEL WORKS, SHEFFIELD.

77 Brightside Steel Works in 1910. This photograph shows the railway line into the works together with the cementation and crucible furnaces in the Top Works and the reality of the scene is in sharp contrast to the idealised engravings shown previously. The long building on the left is typical of those erected in many works from about 1880 onwards to house open-hearth furnaces (in which case tall chimneys should be associated with them), large forges, cogging mills or machine shops; it is, however, not easy to recognise activities from chimneys any more.

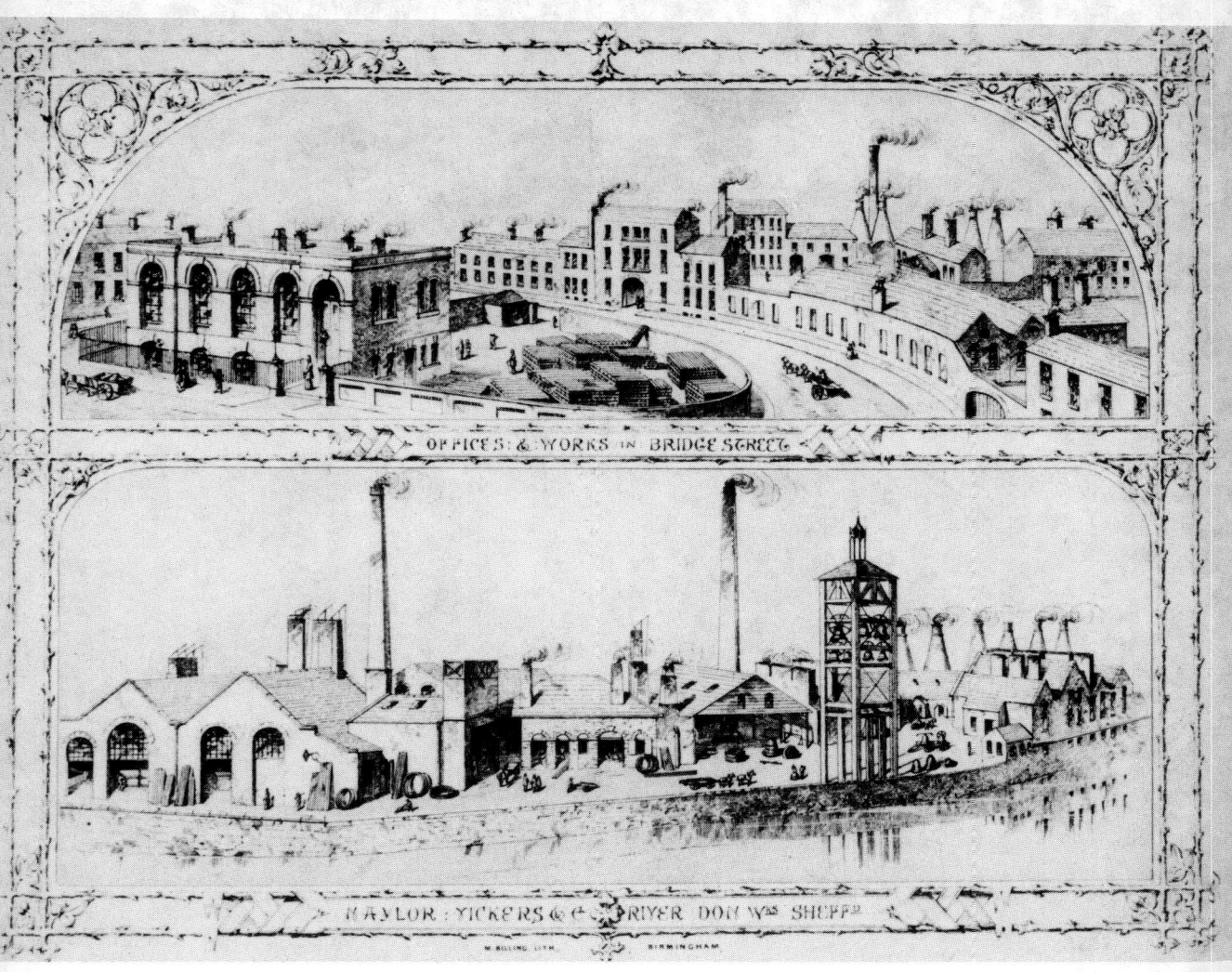

79 Naylor, Vickers and Company in 1858. This print shows, in addition to the Bridge Street Works, the River Don Works. Lest this should be confusing, it should be clearly stated that this works was generally known as Millsands, admittedly on the River Don, and not the more famous River Don Works, down Brightside Lane, which was built between 1862 and 1864. Here, at the older works of this name, we see crucible furnaces, cementation furnaces and forge chimneys, with further cementation furnaces at the Bridge Street works. Note particularly the tall structure, quite clearly a test tower for cast-steel bells (see Figs 54 and 55).

78 (*opposite, bottom*) **William Jessop and Sons, Park Steel Works, in 1879.** The Brightside Works were not the sum total of the Jessop organisation. They also owned Soho Rolling Mills in the town, together with the Park Works in Blast Lane, shown here. This works also had its cementation and crucible furnaces. It was demolished in 1898 to make room for the extension of the railway goods depot.

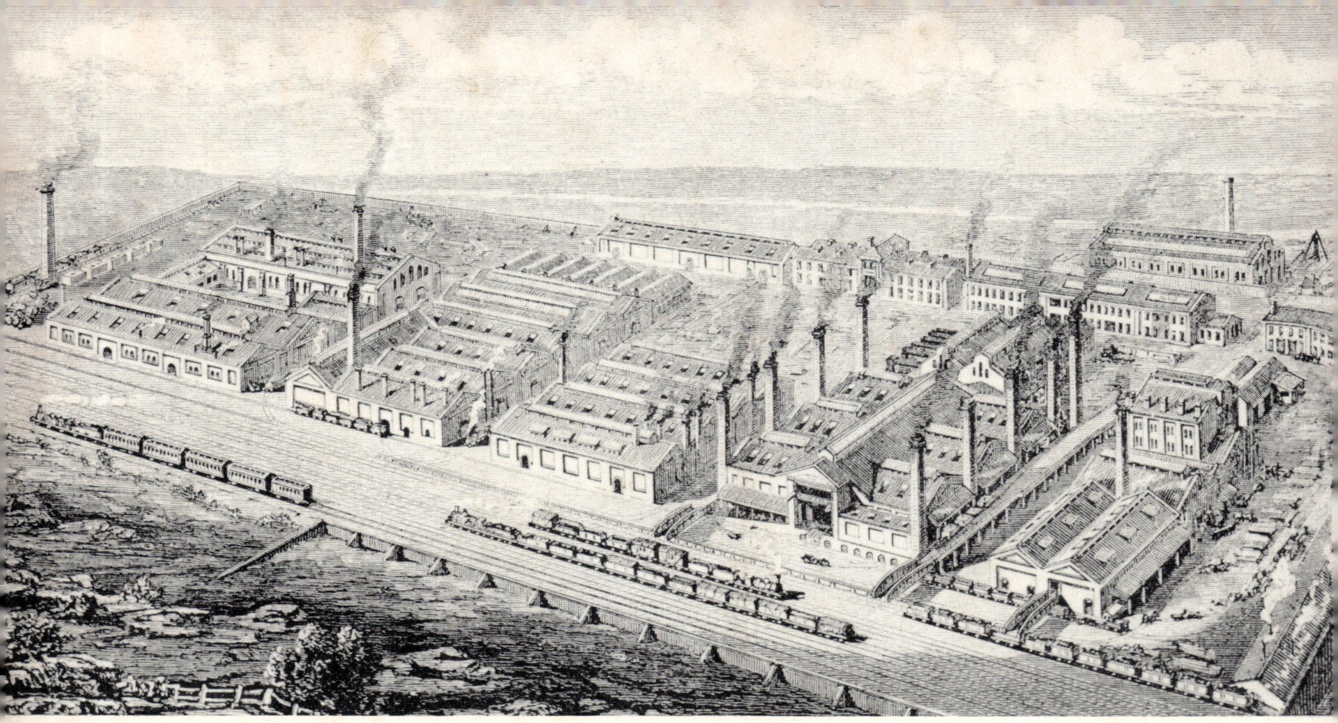

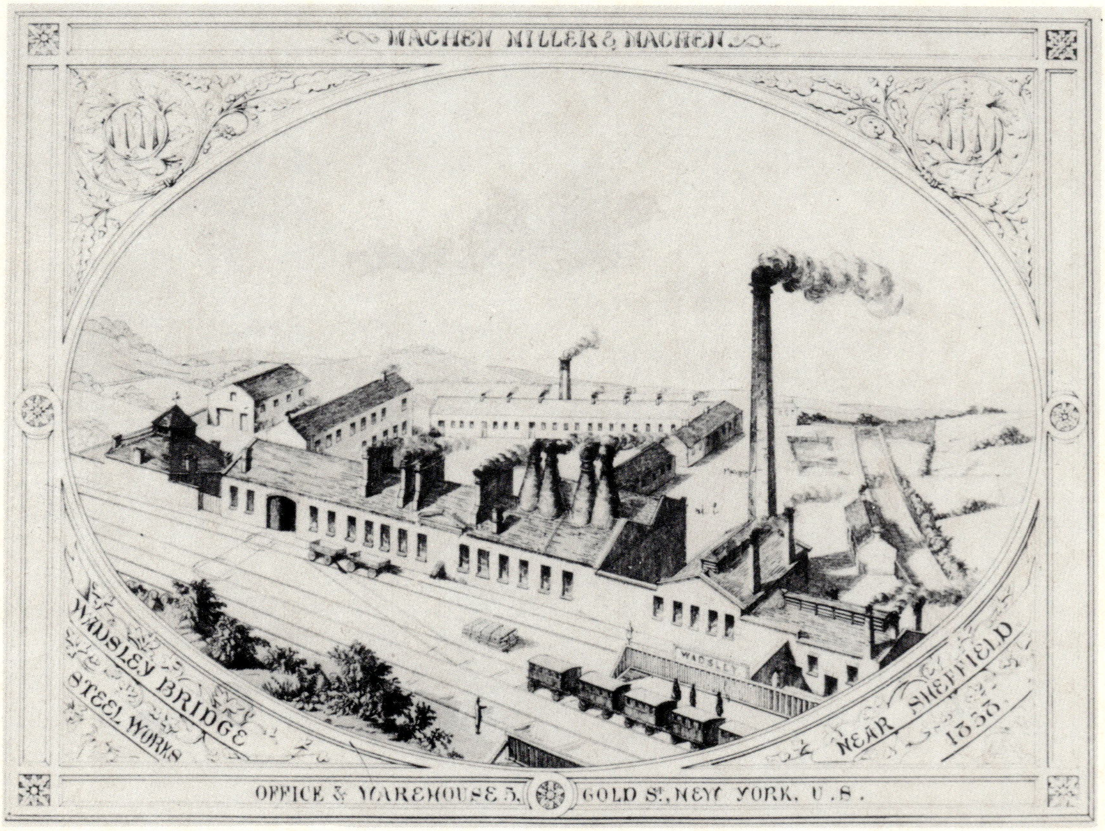

WADSLEY BRIDGE STEEL WORKS
NEAR SHEFFIELD
1855
OFFICE & WAREHOUSE 5, GOLD St, NEW YORK, U.S.

80 (*opposite, top*) **River Don Works in 1879.** This is the new works, down Brightside Lane, the company having changed its name to Vickers, Sons and Company in the interim. Here was the largest conglomeration of crucible steelmelting capacity in the world, capable of producing some 15,000 tons per annum at peak output. Note the complete absence of cementation furnace chimneys; presumably they were carrying out the method set out in the patent of Thomas Vickers, using charges of Swedish bar-iron mixed with the appropriate amounts of Swedish white cast-iron (see Fig 42).

81 (*opposite, bottom*) **Wadsley Bridge Steel Works in 1858.** This works is of some interest in that it was built right out in the country at the crossing of the Sheffield to Penistone road and the Sheffield, Manchester and Lincolnshire Railway, thus relying heavily on rail transport for its raw materials and despatch of its products. It also exhibits an unusual combination of cementation and crucible furnaces within one block. Even such a firm could boast an office in New York, however.

82 J. H. Andrew and Company, Toledo Steel Works, in 1879. Here was a works situated on the River Don, to the north-west of the town. There are numerous crucible furnaces – in the block alongside the street and also along the river bank at the left – but no cementation furnaces. The large block at the right-hand side houses forges and mills, to judge by the chimney structures.

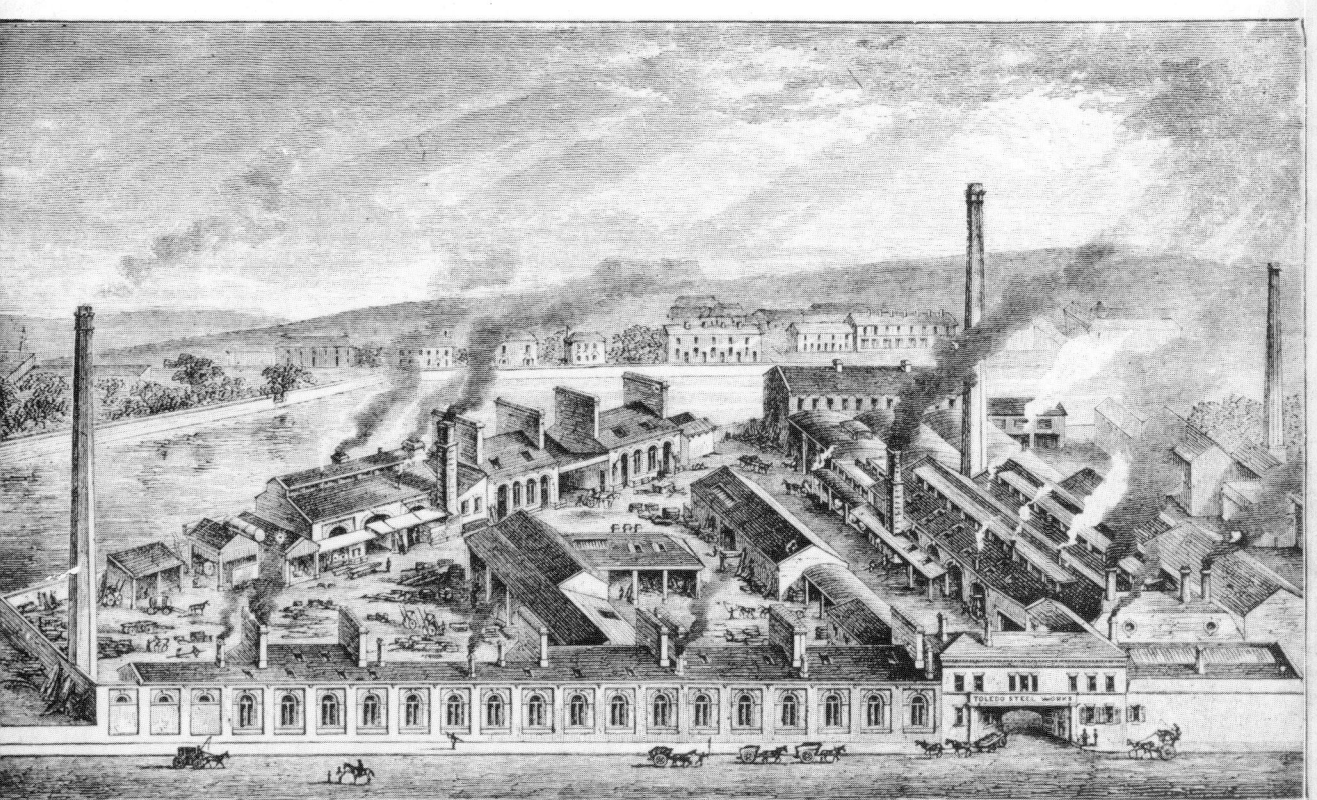

83, 84 and 85 Cocker Brothers Works in 1862. This firm was famous for its steel wire. The three departments, fairly widely separated, all included cementation furnaces. Melting of the steel was carried out at the Navigation Works, the forging and rolling 'out in the country' at Wardsend, while the wire drawing itself was done at Nursery Street, in the centre of the town.

86 Samuel Fox, Stocksbridge, in 1862. This establishment really was out in the country, some ten miles from the centre of Sheffield. Its specialities at this time seem to have been umbrella frames and crinoline wire. We can observe two cementation furnaces and a crucible melting shop occupying a very small area of the works.

87 Davy Brothers, Park Ironworks, in 1879. Set among the many steelworks was the Park Ironworks, founded about 1845 and specialising in engineering and the manufacture of steelworks plant. Here we can see the works with the River Don in the foreground, the Midland Railway to the right and the Great Central Railway in the background.

88 Davy Brothers, Park Ironworks, about 1900. This old photograph shows the entrance to the works on Leveson Street. The works later passed into the ownership of Charles Cammell and Company. (See also Figs 127 and 128).

Later Steelmaking Processes
Bessemer Steel

89 Bessemer and Company in 1879. Bessemer's invention was announced in 1856, but Sheffield was slow to take up the method and Henry Bessemer therefore set up his own works in Carlisle Street in 1858-9. Soon John Brown, Charles Cammell and several others became interested and numerous Bessemer converters appeared within the Sheffield steelworks over the next ten years or so. Here in Bessemer's own works we can observe a vessel of completed steel being tipped to fill a ladle, later to be used to cast ingots in the moulds set up within the pit. By this method, one vessel would eventually be capable of producing some 20 to 30 tons of steel in less than half an hour; compare this with the use of crucible pots containing 50 to 60lb of metal and requiring three to four hours for melting!

90 Charles Cammell and Company, 1897. This is a photograph of a very similar establishment to Bessemer's own, with two Bessemer converters, moulds in the casting pit, with ladle and casting crane.

91 John Brown and Company, 1889. This is a souvenir of the visit made by the Shah of Persia to Sheffield. Here at the Atlas Works is seen a Bessemer converter in blast; the ejection of flames and sparks of burning metal was one of the most spectacular sights in a steelworks. The implication in the original caption to this drawing is that the operation is within a foundry and it could be that the items surrounding the furnace are the artist's impression of sand moulds in a circular casting pit.

Open-Hearth Steel

92 and 93 *(opposite)* **Parkgate Works, about 1900.** This was originally an ironworks, using the puddling process to make wrought iron from pig iron produced in its own blast furnaces. It was converted late in the nineteenth century to a steelworks based on the Basic Open-Hearth process. The pig iron used at the works originally came from the Holmes blast furnaces (which may be seen in the background to Fig 15), but by 1860 more modern and larger blast-furnaces had been erected at Parkgate itself, some seven miles from Sheffield; when the change to steel came, these were capable of supplying the hot metal required. It will be noted that a full range of mills was available for producing plate, sheet, bar and rod from the ingots coming from the melting shop.

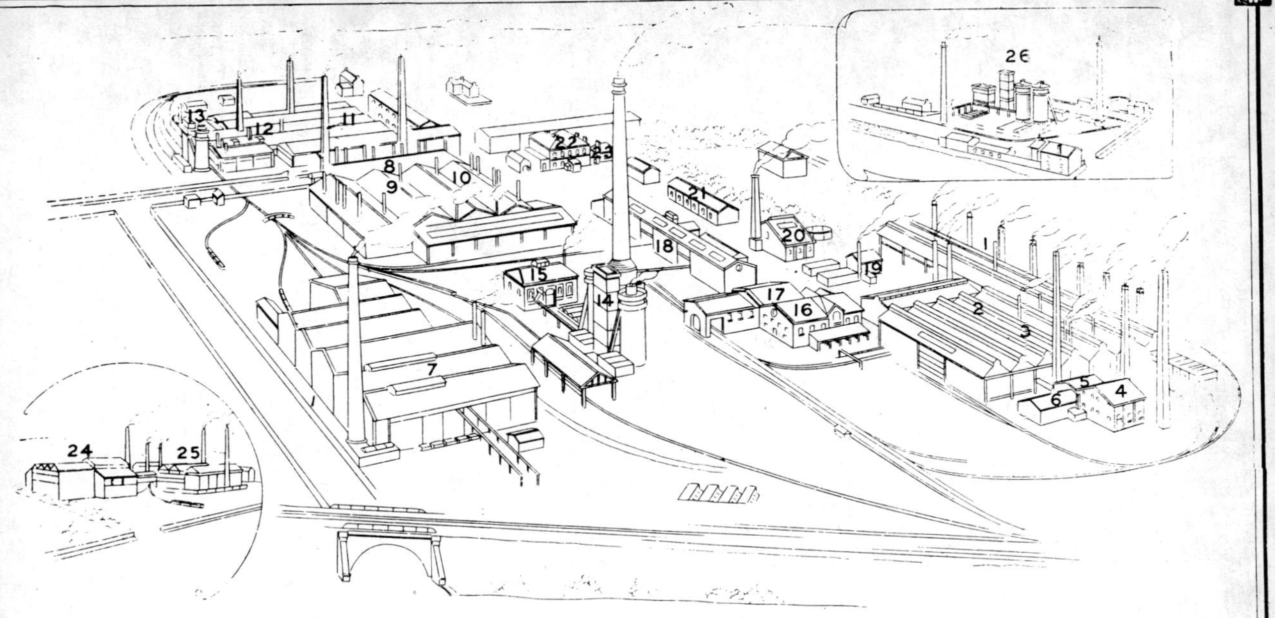

KEY TO ENGRAVING

1. Siemens Furnace Melting Shop.
2. Cogging Mill and Billeting Mill.
3. Steel Plate Mill.
4. Electric Power House.
5. Hydraulic and Pneumatic Power House.
6. Test House.
7. 20 in., 16 in. and 9 in. Bar Mills.
8. No. 3 Forge.
9. No. 1 Plate Mill.
10. No. 2 Plate Mill.
11. No. 1 Forge.
12. Sheet Mill.
13. No. 1 Blast Furnace.
14. No. 2 and 3 Blast Furnaces.
15. Blowing Engine House.
16. Boiler Shop.
17. Iron Foundry.
18. Fitting and Machine Shop.
19. Brick Works.
20. Gas Works.
21. Stables.
22. General Offices.
23. Laboratory.
24. Basic Slag Manure Works.
25. Rail Mill.
26. Holmes Blast Furnaces.

94 Fettling an Open Hearth furnace.
The series of illustrations which follows come from photographs taken at Parkgate as late as 1924 but, apart from a few details, they represent the practice as it had been carried on for the previous 30 years. To 'fettle' is to put in order; here the term implies repairing any damage to the furnace lining which may have occured during the previous heat and consists of the placing of suitable refractory material (in this case carefully graded grains of dolomite) to fill any holes, prior to putting in a further metal charge.

95 Charging molten blast-furnace metal. This shows the transfer of a ladleful, about 25 tons, of molten metal from the blast furnace to the fettled Open Hearth furnace, which could produce up to 70 tons of steel in about ten hours.

96 Charging scrap to the furnace. The other part of the metal charge consisted of scrap. To produce the necessary slag, lime was added and when the charge was fully molten the carbon would be 'boiled out' by the addition of iron ore. The so-called 'boiling' was, in fact, agitation of the metal and slag by the bubbles of carbon monoxide produced by the oxidation of the carbon present in the molten metal, the source of the oxygen being the iron ore added to the slag. Since the slag was basic, that is rich in lime, it was capable of taking up both those unwanted elements, sulphur and phosphorus. The furnace was fired with producer gas which mixed with preheated air from the Siemens regenerative chambers situated beneath the furnace; the furnace principle should be sought in any steelmaking text-book by the interested reader.

97 Relining the ladle. The metal from the furnace, when converted into steel of the desired composition, was run out into a ladle, an iron pot clad internally with refractory bricks so as to hold the liquid metal with safety and at the same time retain its heat. Here we see such a vessel being relined with refractories. Some idea of the size may be obtained by comparison with the men performing the task. The stopper-rod system, two units being fitted, can also be seen to the left of the picture, this mechanism being used to control the flow of liquid metal through the nozzle after the fitting of the stopper rod (see Fig 100).

98 and 99 Tapping the steel into the ladle. Here we have two views of the metal running from the furnace into the ladle. The movement of the ladle car by electric motors is a twentieth-century variant of the original. Most ladles were suspended from an overhead crane by wire ropes in the earlier days (and, indeed, still are in a number of shops).

100 Casting the ingots. The downward movement of the stopper lever raised the refractory-sleeved stopper rod from the nozzle seat and metal then flowed out in a stream from the nozzle into the ingot mould waiting below; raising the end of the lever checked the flow before moving on to the next mould. The second, longer lever from the back of the ladle controlled a second stopper, for use in case there was any mishap to the first one.

101 Ingots of Parkgate steel. When solid, the ingots were stripped from the moulds. Here is a group of ingots suitable for rolling in the cogging mill to produce billet, probably weighing two to three tons each. Note the significant difference in this straightforward method of producing bulk steel with the earlier crucible method. Nevertheless, crucible steel, although much more costly, continued in demand well into the twentieth century.

The Forging of Steel

102 Waterwheel. The earliest form of steel forge was based on water power, deriving from the still older 'helve' hammer used in the iron trade, as exemplified by those still preserved at Wortley Top Forge, some eight miles from Sheffield on the Upper Don. This illustration shows a waterwheel which was in operation in the Jessop Brightside works until the end of the nineteenth century.

103 Claywheels Forge about 1900. This forge, near Wadsley Bridge, was worked by Thomas Firth and Sons throughout the latter half of the nineteenth century, subsequently being taken over as a scythe forge.

CLAY WHEEL FORGE.
1848 – 1905.

104 *(opposite)* **Claywheels Forge about 1870.** This gives an interesting comparison; the artist can show more of the site than a photographer can and when the reporting is accurate the result is most informative. The photograph shows two waterwheels with a rather indeterminate structure between them; the artist shows all four wheels and it also becomes clear that each has its own 'pentrough' or water reservoir alongside, the flow of water from this to the upper part of the wheel, and thus its speed of rotation, being controlled by a shuttle operated by a lever from inside the building alongside the hammer. The right-hand-side wheels seem to be bigger in diameter, and thus more powerful, than the left-hand pair. The water supply came from the dam at the rear which can just be discerned through the smoke.

105 Interior of Claywheels Forge about 1940. This shows the forging of a scythe in a shop very similar to the one preserved at the Abbeydale Hamlet (see Figs 158-160). In the days when it was worked by Firths these hammers were used to cog down the small crucible-steel ingots into billets (using the larger unit on the left) and these billets into bars at the smaller unit, here in operation on a scythe. Note the swinging seat for manoeuvring the blade beneath the hammer. Claywheels Forge has since been completely demolished.

106 Steel Tilting in 1862. Here we are in the Sanderson Brothers' works at Attercliffe and have the next development stage: the bringing together of a number of the old-type hammers into one shop. Obviously these could be run by a number of waterwheels, but it is believed that this particular shop was powered by a steam engine at this time. The 'scissors' in the foreground, used for cutting the bars, were driven from the main motive-power by a cam system. A similar arrangement can still be seen at Abbeydale.

107 Tilt Hammers at Brightside Works about 1900. The use of water power did, however, continue elsewhere. These hammers in the Jessop works continued to be used at least to the end of the century and are now preserved at the entrance to the Abbeydale Hamlet. The three hammers were all driven from a common shaft powered by a single waterwheel.

108 Steam Hammers at the Atlas Works in 1865. This, and the next illustration, must be amongst the earliest photographs taken within a works. The steam hammer was invented by Nasmyth in 1842 and the first hammers installed soon after only had a force on the block of a few hundredweights. They were still known as tilt hammers, even into the twentieth century! Over a period of twenty years the steam hammer developed and eventually monsters which shook the ground and upset the sleep of whole neighbourhoods were installed. The two hammers shown here are not identified but would appear to be about 10 ton units; such machines had a long life and many have survived until recent times. Note the crane at the extreme left – it also appears in the next illustration.

109 and 110 25 Ton hammer at Atlas Works in 1865. It is unusual to find both an engineering drawing and an actual photograph of a piece of plant at such an early date, but here we have evidence of a monster hammer built for John Brown by the Bradford engineering firm of Thwaites and Carbutt. The great weight of the ingots which had to be handled involved the use of teams of men operating levers to move them back and forth under the hammer.

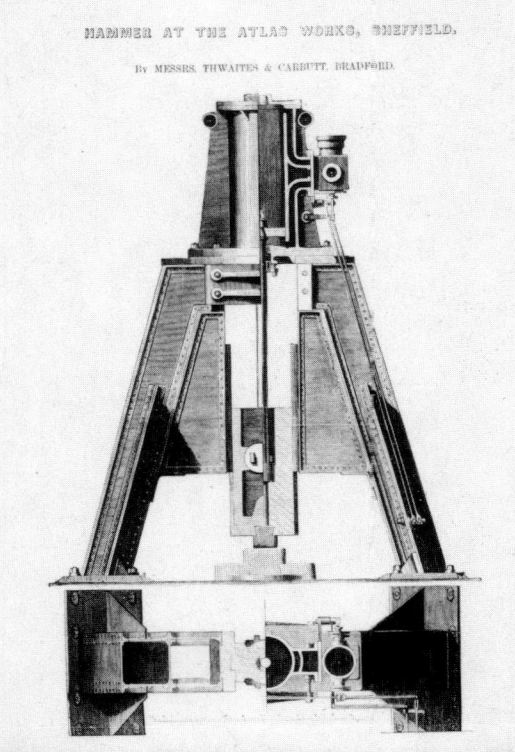

111 Large hammers at the River Don Works in 1898. Here we can see two hammers of comparable size to the Atlas hammer at the Vickers, Son and Maxim works. Note the similar number of men in the forging team to that in the previous example from thirty years earlier.

112 Hydraulic Press in 1898. This was the latest addition to the forge equipment in the nineteenth century. Again at the River Don works, we see a hydraulic press designed to exert a force of 8,000 tons on the ram, capable of forging the ever-larger ingots then being made, these having proved to be a difficult proposition for even the largest hammers. The slab seen under the press is stated to have been 18in thick armour-plate.

Rolling of Steel

113 Water-powered bar mill, dated 1845. This mill was not in Sheffield, but in 'New Sheffield', the name originally given by John Spencer to his steelworks at Newburn on Tyne, where the traditional Sheffield steelmaking methods were carried out. It is included here in default of any other known illustration showing a mill coupled to a waterwheel, although several are known to have existed in the Sheffield area. This is a direct descendent of the 'grooved roll' mill invented by Cort alongside his puddling process in 1784 and some remains from an early mill of this kind can still be seen preserved at Wortley Top Forge. Each pair of working rolls had semi-circular grooves, reducing in diameter in stages along the barrel, so that a larger bar could be reduced gradually in section by passing it through one hole after the other, until the desired size was reached; this might involve passing to a second 'stand' with a set of smaller holes in the rolls. Here we have four working-stands with coupling boxes between them.

114 Steel rolling mills at the Cyclops Works in 1862. These mills are essentially similar to those in the last illustration, but are driven by a pair of steam engines with rocking beams, gearing and a flywheel. Note the length of the shop and the overhead glazing. There are the reheating furnaces on the left-hand side, and a pair of 'scissors', while the rolled bars are dealt with on the right.

115 Rod mill at Brightside Works about 1900. A still later type of mill, with a more modern vertical steam-engine, but still recognisably the same type of equipment.

116 Fourteen inch rolling mill, Hobson, Seaman and Company, 1897. A somewhat larger mill; the term 'fourteen inch' relates to the diameter of the barrel of the working rolls. Note in both this and the last illustration the various spare pairs of rolls available for putting into the mill to roll various specific types of sections of bar.

117 Cogging mill at Parkgate in 1924. A 'cogging mill' is used for converting ingots into billets or slabs for further processing down in smaller mills. Such mills nowadays have rolls up to 40in or so in diameter; in the nineteenth century 24 to 30in mills were more usual and this is one of this general type, installed to handle the ingots which we have seen earlier (see Fig 101).

118 Reheating furnace at Parkgate. All the mills we have mentioned required the steel, whether it was ingot, billet or slab, to be heated to a suitable temperature, and moreover a uniform temperature, through the mass, to enable it to be deformed in the desired manner by the roll pressure available. All mills, therefore, were provided with furnaces for heating the steel. In the case of ingots, which being larger required more time for heating through, this was done in 'soaking pits', usually fired with producer gas. Here we can observe the withdrawal of an ingot from such a pit prior to transfer to the cogging mill.

119 Sheet rolling at Parkgate. To produce sheet, a rectangular ingot would first be rolled to a slab, typically some such section as 14in × 3in. This slab would be cut into pieces, these reheated and rolled down further to thickish plates, known in the trade as 'moulds' or 'sheet bar', which would then be reheated and rolled into sheets in a mill such as this one. The final thickness was adjusted by use of the screwdown equipment on each stand, which varied the gap between the rolls.

120 Plate rolling at Parkgate. To produce plate the slab would be rolled from the ingot, cut to the appropriate weight and reheated. It would then be rolled first to the required width, then turned through 90 degrees and rolled to the required thickness, which automatically gave the length if the weight calculation was correct! Sheet and plate mills had what were essentially plain barrel-rolls with as good a finish as practicable to impart a good surface to the product.

121 *(opposite, top)* **Armour-plate rolling at the Cyclops Works in 1879.** Here we see a much larger scale of operation. Rolls up to four feet in diameter and with barrel lengths over six feet in length handled slabs (or slab-shaped ingots) and rolled them into plates anything from an inch or two in thickness to well over a foot; the individual pieces could weigh several tons.

122 *(opposite, bottom)* **Armour-plate rolling at Atlas Works in 1863.** This famous drawing of the withdrawal of the slab from the furnace by some forty men is indicative of the situation in the early days of bulk-steel production. Although in this case it seems more than likely that the material being man handled was wrought iron, the plant was sufficiently robust to handle steel equally well. The need for teamwork in this larger scale operation is again quite obvious.

123 Armour-plate rolling at the River Don Works in 1898. A more modern mill, with more mechanical handling, at the end of the century, here seen rolling an 18in thick plate. This mill, incidentally, still stands and has rolled a variety of materials, including stainless steel, in recent years.

124 Armour-plate rolling at Cyclops Works in 1897. This depicts an exhibition of armour-plate rolling arranged for the benefit of Queen Victoria on her visit to Sheffield in her Diamond Jubilee year. And it was a most spectacular process. The rollers threw bundles of heather on to the plate as it entered the rolls. With the heat and the pressure, the heather burned and exploded in a fusillade of loud reports as the sap was suddenly converted into steam. This effectively blew the scale off the surface of the plate and prevented it being rolled in. It may well be that Queen Victoria was amazed, if not amused, on this occasion!

125 *(opposite, top)* **Rolled tyre-rings at Steel, Peech and Tozer's Works, about 1920.** A final example of the shaping of steel by rolling is the production of 'weldless' rings and railway-wheel tyres. Such items were first produced in Sheffield some time just before 1870, by the rolling of a forged ring-blank on a so-called 'tyre mill'. This reduced the section and thus increased the circumference of the ring as the work proceeded; it was fascinating to see the ring growing in size and moving upwards and outwards from the fixed pair of rolls in the mill. By using specially-shaped rolls complex sections could be produced, such as actual tyre rings. Here we have either a ring of about 6ft in diameter or three very short men!

Heavy Engineering

126 Gun shop at River Don Works about 1900. A speciality of the Vickers works was the production of guns. In addition to the melting and forging equipment, therefore, there had to be provided the necessary heat-treatment facilities as well as extensive machine-shops to carry out the boring of the forgings, the machining of the exterior and finally the wiring of the barrels and the provision of mountings. This shop had the sole function of producing such guns. Similar large machine-shops were set up by Cammells, Firths and Browns; not only were guns and other munitions produced but marine-shafts and cranks, steam boilers and other general engineering items of large size.

127 and 128 Davy Brothers, Park Ironworks, about 1900. The increase in scale of steelmaking operations led to a number of firms specialising in making the necessary plant, gradually developing into large-scale engineering workshops. We have seen in the last few illustrations the large forge equipment and rolling mills; Davy Brothers became an important supplier of such items, developing their old ironworks to meet these needs. These two views illustrate the scale of their operations. In one shop we can see a mill roller-table in course of erection; the other shows one of their general machine shops.

The Lighter Trades

129 Wire drawing, Arthur Lee, about 1900. The drawing of wire was an ancient trade in the valley of the Don; traditionally Wortley Wire Mill was built in 1624. Steel wire from Huntsman's steel was being used for needle manufacture before 1800 and Cocker Brothers were famous for their steel wire (see Figs 83-5). Here we have a later example, with wire being produced from rod by successive passage through a steel plate with various-sized holes in it, known as a wortle plate. One of these can be seen leaning against the bench and another is clamped in the frame on the drawbench, the wire being pulled through the appropriately-sized hole from one drum to another. The original operations of this type were either man-powered or animal-powered; then came the use of waterwheels and later the shafting and gearing would be steam powered, as appears to be the case here.

130 Joseph Rodgers and Sons in 1879. The most famous product of Sheffield's lighter trades was, and still is, cutlery. Originally more-or-less a cottage industry, cutlery making eventually grew in scope until in the middle of the nineteenth century there were several large works such as this. While blades would still be forged by many of the independent 'mesters' and forks would be produced for instance at Shiregreen and Ecclesfield, and bought in by Rodgers, the polishing and grinding, the fitting of handles and so on would be carried out in the central workshops. In this way many patterns of cutlery were made available for sale all over the world. Other larger works would specialise in edge tools, agricultural implements, files and so on.

131 Forging and grinding shop, 1879. Here we have a glimpse into the works of James Howarth and Sons, edge-tool manufacturers, showing a typical arrangement in the lighter trades area.

The following four illustrations are all examples of specialised forging methods designed for the particular purposes in hand, from standard small-hammer techniques to blanking-out operations with dies. This could be called mechanised blacksmithing, carrying on age-old traditions with what was then new equipment.

132 Spade forging

133 Forging garden shears

134 Shovel and fork forging

135 Light tool forging

136 File forge, 1897. Here at the works of Hobson, Seaman and Company in their Hoyle Street file factory we see small forging machines in the background, with some heat-treatment facilities for hardening the files, together with hand cutting of the file teeth, maintaining a tradition which went back over the centuries. Gabriel Jars described the hardening process as he saw it in a Sheffield works in 1774 and it could well have been the same here more than a century later.

137 File cutting shop in 1879. Here we see a row of some thirty file-cutters; note that they are facing the windows, probably arranged to be north facing. This was in William Hall's works in Alma Street. Cutting teeth by hand was practiced for the best-quality files well into the twentieth century, despite very efficient machinery introduced for the purpose from around 1880 onwards. There was, indeed, a long and often bitter controversy as to the relative merits of the two methods.

138 Machine cutting of files in 1897. It seems that Hobson, Seaman and Company backed both horses! We noted hand cutting of files in their shop in Fig 136; here, in their same works, is a batch of file-cutting machines in a shop stated to be 105ft long and 25ft wide.

139 Sheep shear making, about 1900. Sheep shears had been produced in Sheffield since the seventeenth century; it is just conceivable that 'shear-steel' was so-called because of its use for this purpose. The fame of the Sheffield article was widespread (and the author well recalls a pair with a well-known Sheffield name stamped on them, over a hundred years old, being shown to him after a shearing on the Isle of Skye). Here we can see a blanking-press stamping out the form from a thick sheet of steel.

140 Sheep shear grinding. After blanking-out and hardening, the blades required grinding; here we see a small army of grinders occupied on the task.

141 Sheep shear bending. The shears then required bending to make the two blades fit snugly together. A pile of ground blanks prior to bending is being inspected; the bending is then being carried out, to be followed by setting to ensure a good cut.

142, 143 and 144 Grinding. Each trade had its own special grinders. In these three views we can see the operation on larger edge-tools, scissors and pen knives.

144 Grinding, pen knives.

145 Spade and shovel finishing. Each establishment had its own finishing department. Cutlery, scythes, spades and shovels, all had to have handles fitted, had to be polished up and, of course, had to have the maker's name inscribed. Here we see the fitting of handles to spades.

146-154 Illustrations from 'The Sheffield Catalogue', 1909. Here we have a few representative items of the wide variety of Sheffield manufacturers: 'Bills and Slashers', 'Shears', 'Knives for the Kitchen', 'Files', 'Turning Tools for Metal', 'Axes', 'Scissors', 'Pen-knives' and, finally, 'Gentlemen's Tool Cabinets' – would the latter were still available at the same price! Note the catering for regional needs as shown in the differing bills, slashers and axes.

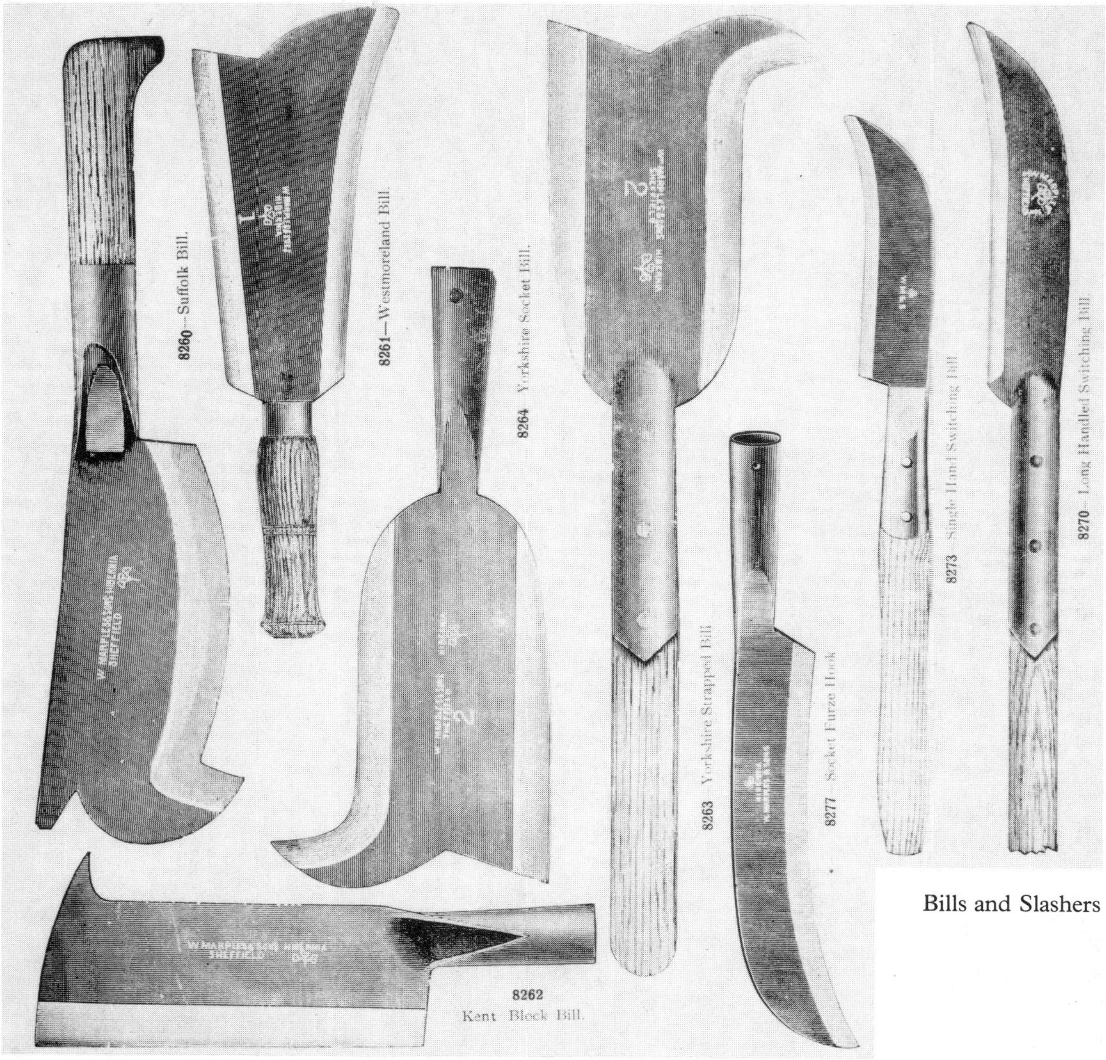

Bills and Slashers

Shears

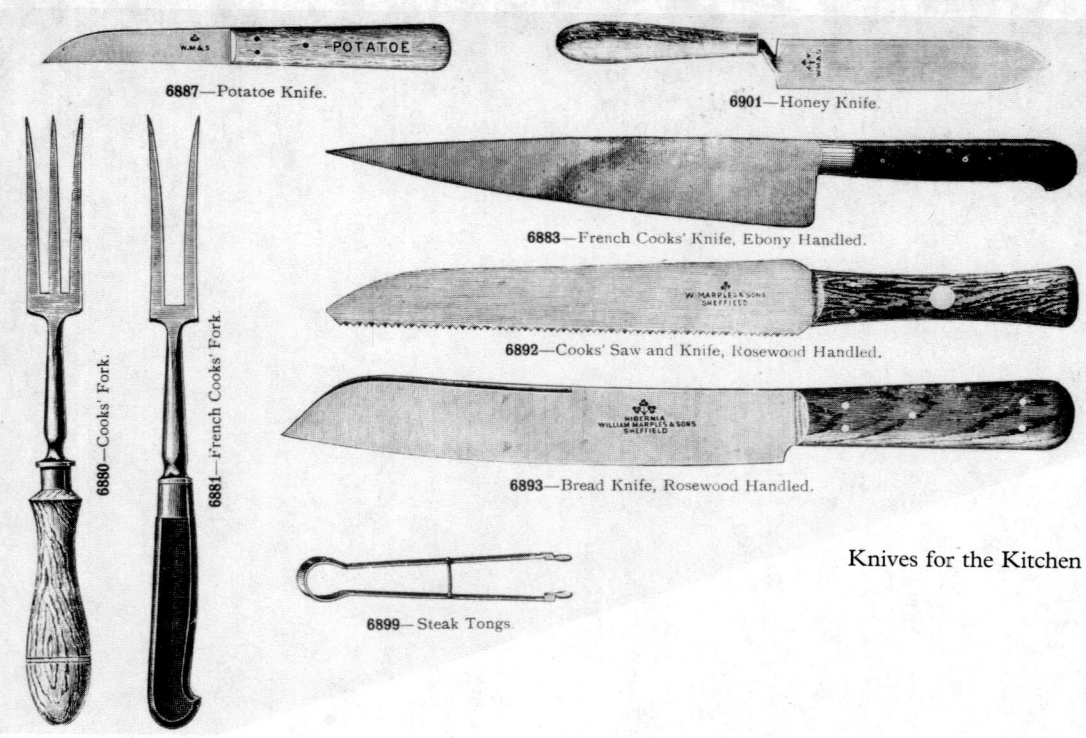

Knives for the Kitchen

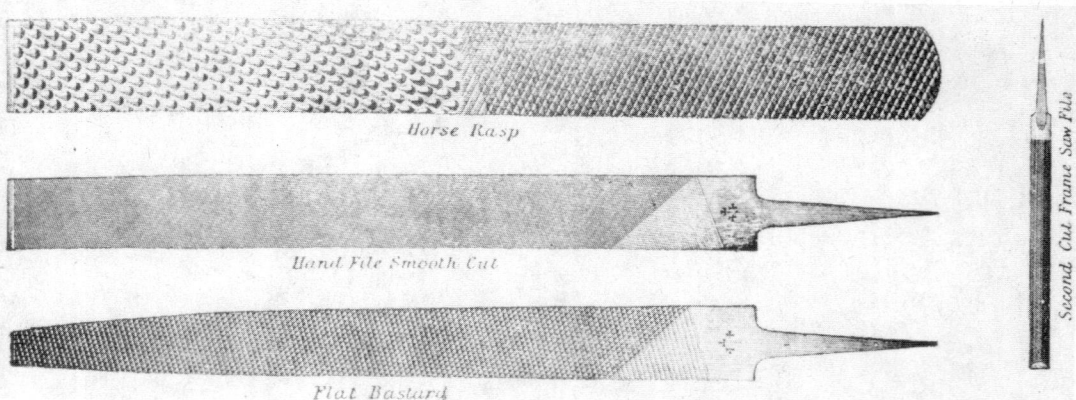

Files

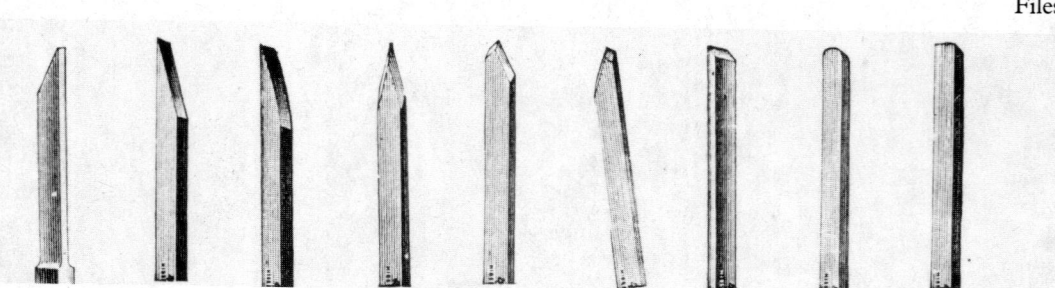

Turning Tools for Metal

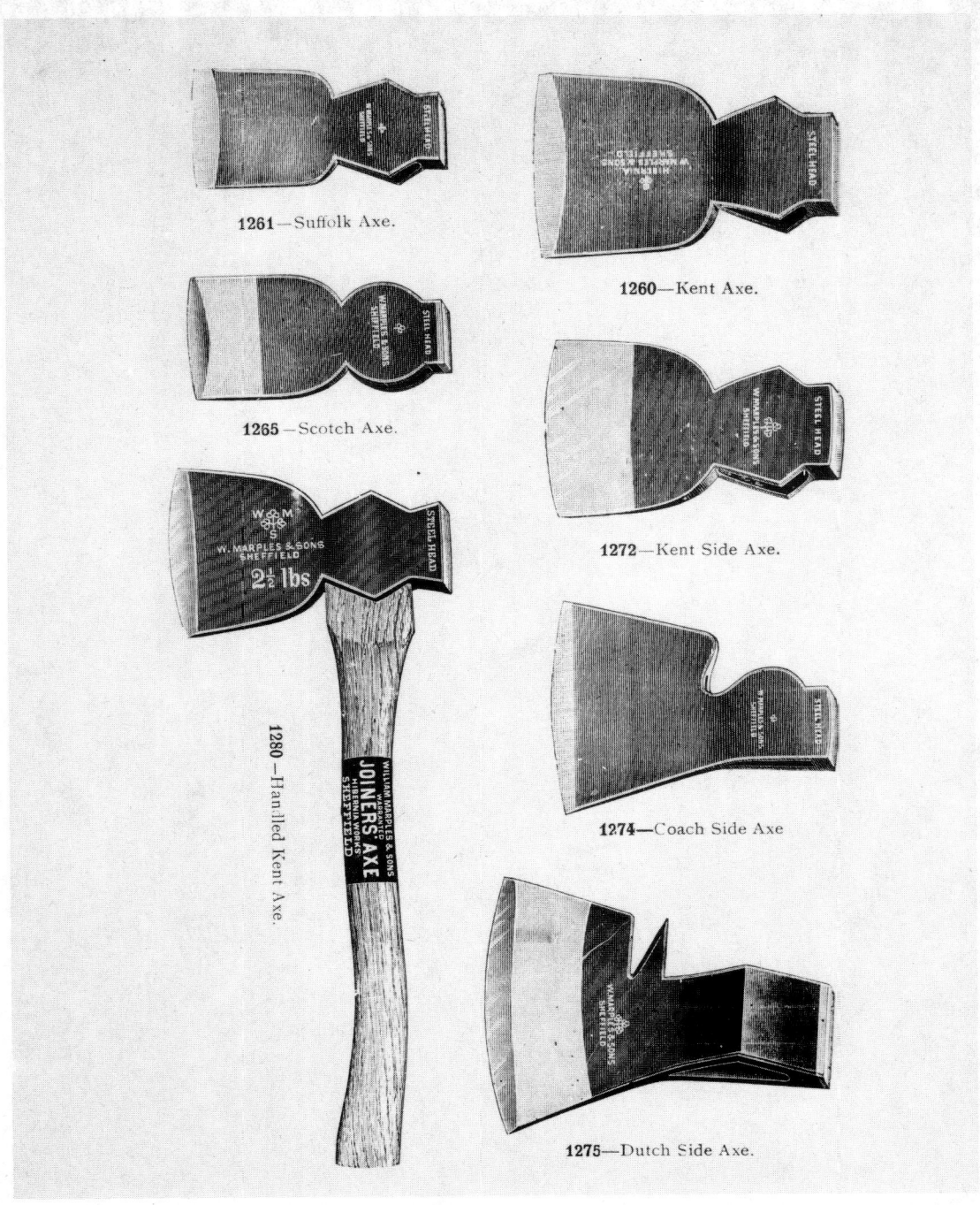

Axes

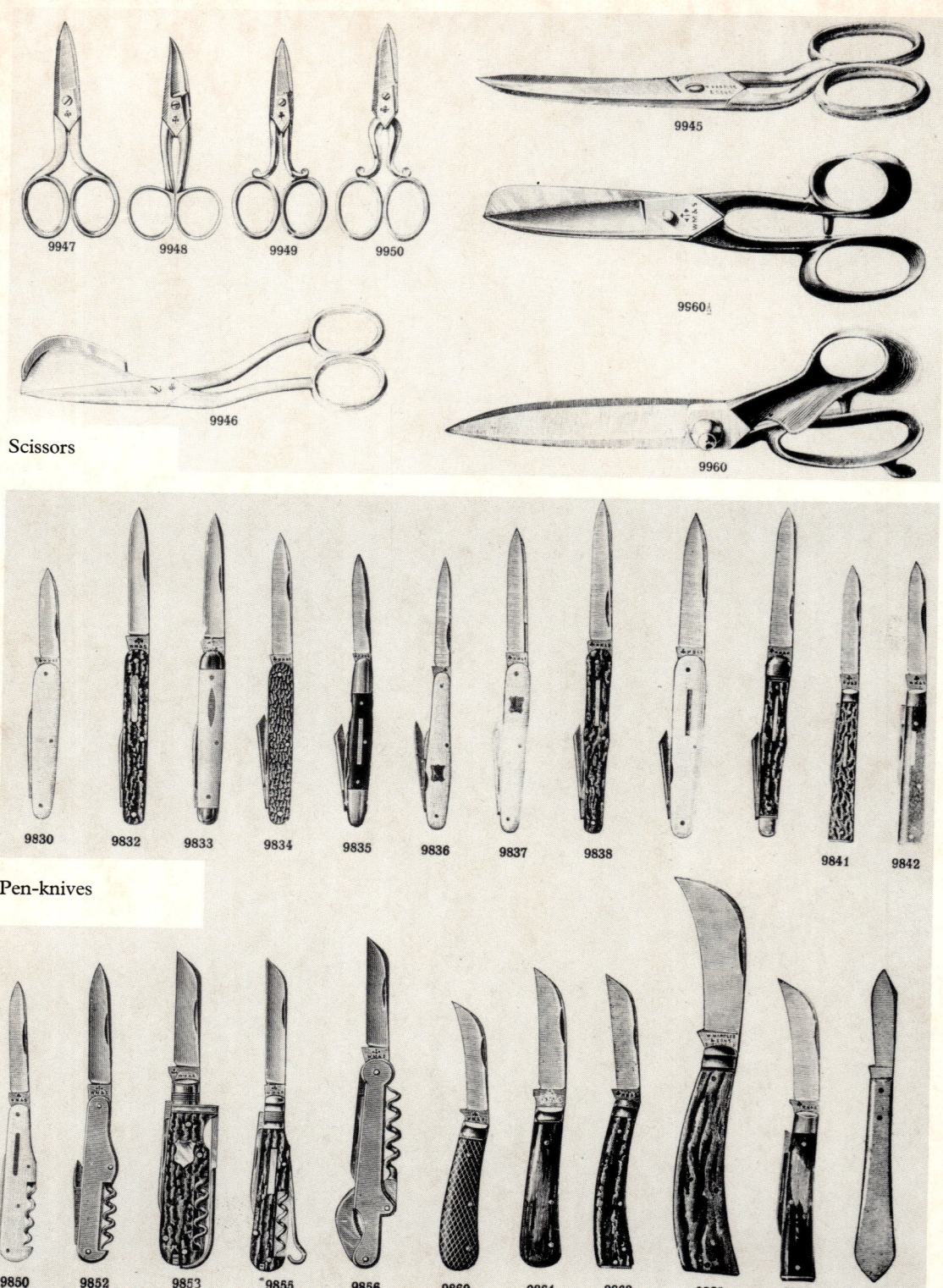

Scissors

Pen-knives

THE "COMPACTUM" PATENT TOOL CABINET.

Made of dark American Walnut with Ash Panels, forming, when closed, a handsome piece of furniture.

The Hardwood Top, is 3ft 6in. long by 1ft. 6in. wide. The Extension Piece, 1ft. 6in. square, can be fitted either in front, for Wood Carving, or at the end, for Carpentry, as shown in the Illustrations.

The Tool Rack can be raised with ease by a special appliance, and when lowered is flush with the top of the Bench.

The Cabinet is furnished complete with full-sized Tools of Best Warranted Quality.

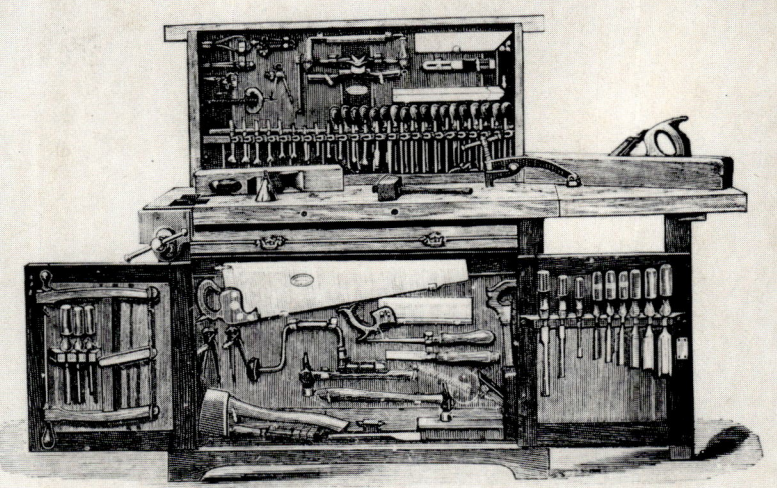

No. 9015½—"Compactum" Cabinet, with instantaneous Vice and Tools
(Fixed as a Carpenter's Bench).

No. 9015—"Compactum" Cabinet, with ordinary Vice and Tools.
(Fixed as a Wood Carver's Bench).

CONTENTS.

Smooth Plane 2¼in., Jack Plane 2in., Try Plane 2½ in., Iron Rabbet Plane, N.P. Ratchet Brace and 24 Bright Bits, 6 Gimlets, 6 Brad Awls, Rosewood Mortise Gauge, Marking Gauge, Cutting Gauge, Try and Mitre Square, Ebony Bevel, C.S. Hand Saw 26in., Brass Back Saw 10in., Compass Saw, Turning Saw Frame, Saw Pad and 6 Blades, Boxwood Spokeshave, Iron Chamfer Shave, 3 Turnscrews assorted, 6 Boxwood Handled B.E. Chisels assorted, 3 Boxwood Handled Gouges, 2 Mortise Chisels, Registered Chisel 1 inch, Draw Knife 6in., Pincers 7in., Pliers 5½in., Cutting Nippers 6in., Wing Compasses 6in., In and Out Callipers, Spring Dividers, Spirit Level, Boxw'd 2 fold Rule, 3 Hammers assorted, Joiner's Mallet, Wood Carver's Mallet, 18 Handled Carving Tools assorted, Carver's Screw, Sharpening Slips, Mitre Block, Bench Holdfast, Handled Axe, Glue Pot and Brush, Stocked Washita Oilstone, Gouge Slip, Screw Wrench, Anvil, Hand Vice, Cold Chisel, Oil Can, 2 Handled Files and 1 Rasp, Star Punch, Handled Scraper, Glass Cutter, Carpenter's Pencil.

No. 9015—"Compactum" Cabinet, fitted with Ordinary Vice and complete with Tools ... £26 0 0

No. 9015½—"Compactum" Cabinet, fitted with Instantaneous Vice and complete with Tools ... £26 12 0

No. 9016—Empty Cabinet fitted with Ordinary Vice, but without Tools or Clips... £13 16 0

No. 9016½—Empty Cabinet fitted with Instantaneous Vice, but without Tools or Clips... £14 8 0

No 9018—Wood Carvers' Stool .... 17/ each

The Remaining Heritage

155, 156 and 157 It has already been remarked that there are few remains of the old furnaces. Here we show three legacies from the steelmaking past. Remembering that the life of a crucible was only a single day and that thousands a year were discarded, it is not surprising that some were incorporated as building materials, as decorative wall toppings or, in some cases, as the main constructional material itself. The third illustration shows the use of another waste product – the discarded crust from the top of the cementation chests, known locally as 'crozzle'. It breaks into rather angular chunks with sharp edges, just the thing for deterring young climbers from scaling walls!

Abbeydale Hamlet. One of the main attractions in Sheffield to those interested in the older steelmaking traditions is the Abbeydale Hamlet. This became a scythe works in the early nineteenth century, with its own facilities for melting steel (although it had to buy in the blister steel for charging to the crucibles), for forging it into bar, for welding this to iron and forging the whole to produce a scythe blade, which was then hardened, ground, set into its handle on the premises and marketed from there. The preservation of these works really came about because of the single-mindedness of a small band of enthusiasts, originally calling themselves the Society for the Preservation of Old Sheffield Tools (later to become the Sheffield Trades Historical Society, which is still active). When they had made quite sure that the premises would not be demolished, the task of restoring them fell to another body, now well-known as the Council for the Conservation of Sheffield Antiquities, which was also responsible for the restoration of Shepherd Wheel, an old water-powered grinding 'hull'.

158 A view of the crucible-steel melting shop (left) with its characteristic chimney, hardening shop (centre) and tilt shop (right). The large mill pool in the background supplied water to turn the waterwheels (just off the photograph to the right of the tilt shop), which powered the tilt hammers and the forge air-blowing machine.

159 The interior of the melting shop showing the furnace holes in the floor, crucibles drying on the shelves to the left and used ones on the right. Tongs for lifting the crucibles, testing bars, coke basket and fork, charging funnel and pan, and ingot moulds can also be seen.

160 *(below)* The tilting shop with two hammers lifted by a series of small cams on the waterwheel-driven shaft. On the left are shears activated by a single large cam on the main shaft, and used for cutting bar to length. Note also the swinging seat and the lever in the centre for controlling the flow of water onto the waterwheel. The large gear on the main waterwheel-shaft has wooden teeth held in place with iron wedges.

Bibliography

The following works were consulted during the preparation of this volume:

G. Broling, *Anteckningar under en Resa i England, 1797-1799* (Stockholm, 1804)
Charles Cammell and Company, Brochure, 1898
D. Carnegie and S. Gladwin, *Liquid Steel: Its Manufacture and Cost* (1913)
J. Holland, *Sheffield and its Neighbourhood* (Sheffield, 1865)
J. Horsfall, *The Ironmasters of Penns* (1971)
J. Hunter, *History of Hallamshire*, ed A. Gatty (1869)
Benjamin Huntsman and Company, Brochure, 1930
G. Jars, *Voyages Metallurgiques* (Lyons, 1774)
William Jessop and Sons, *Visit to a Sheffield Steelworks* (1913)
William Marples Ltd, *Sheffield List* (1909)
A. McPhee, *The Growth of the Cutlery and Allied Trades to 1814* (unpublished)
Pawson and Brailsford, *The Sheffield Illustrated Guide* (1862 and 1879 editions)
F. le Play, 'Memoire sur la Fabrication de l'Acier en Yorkshire', *Annales des Mines*, 4me Serie, Tome **III** (1843)
S. Pollard, *Three Centuries of Sheffield Steel* (1954)
C. Sahlin, 'De Svenska Degelstalsverken', *Med Hammare och Fackla* (1932)
J. D. Scott, *Vickers – A History* (1962)
H. Seebohm, *On the Manufacture of Cast Steel* (published privately, 1869)
Sheffield and Rotherham Illustrated (1897)
John Spencer and Company, Centenary Brochure (1910)
Vickers, Son and Maxim, Brochure (1897)
Ward and Payne Ltd, *Sheffield List* (1911)

To the reader wishing to know more of the technical processes involved, it is suggested that he should refer to some of the older textbooks and in particular to:

J. Percy, *Metallurgy: Iron and Steel* (1864)
J. S. Jeans, *Steel: Its History, Manufacture, Properties and Uses* (1880)
W. H. Greenwood, *Iron and Steel* (1884)

These are available for reference at most large libraries; alternatively, there are two information leaflets issued by Sheffield City Museums:

No. 7: *On the Origins of the British Steel Industry*
No. 8: *Crucible Steel Manufacture*

For general information on Sheffield, its growth and its history the following volumes will prove of interest:

R. E. Leader, *Sheffield in the Eighteenth Century* (1901)
J. H. Stainton, *The Making of Sheffield* (1924)
Mary Walton, *Sheffield: Its Story and Its Achievements* (1952)
British Association, *Sheffield and Its Region* (1956)

Readers may also note that the 1862 edition of the Pawson and Brailsford *Sheffield Illustrated Guide* has recently been reprinted in facsimile (SR Publishers, 1971).

Index

Abbeydale Hamlet, 13, 40, 81, 82, 108, 109
Aetna Works, 56, 60
Alma Street, 97
Anderson, Bengt Quist, 36
Andrews, J. H. & Co, 69
Anglo-Saxons, 9
armour-plate mill, 89, 90
Atlas Works, 54, 56, 58, 60, 74, 83, 89
Attercliffe, 11, 18, 36, 56, 82
axes, 102

Bank Furnace, 11
bar iron, 11, 30, 32, 34, 35, 42
bar mill, 85, 86
Barrow in Furness, 14
Bawtry, 12
bell pits, 8, 10
bellows, 10
Bessemer, Henry, 14, 73
Beverley, 9
bills, 101
Birmingham, 9, 25
Blackburn Brook, 8
blast furnace, 10
Blast Lane, 67
Blonk Tilt, 18
bloomery process, 10
bole hills, 10
Bower Spring Furnace, 26, 28, 62
Bridge Street Works, 67
Bridgehouses Station, 20
Brightside Lane, 67, 69
Brightside Works, 39, 64, 66, 82, 86
Brown, John, 13, 14, 54, 56, 58, 60, 73
Brown's Bridge, 58, 60

Cammell, Charles, 13, 14, 60, 61, 62, 72, 73, 85, 89, 90
Cammell's Bridge, 60, 61
canal (Don Navigation), 12, 18, 20, 28, 30
canal basin, 20, 32, 62
Carlisle Street, 60, 61
cast iron, 10
chafery, 11
Chapeltown, 10
charcoal, 8, 10, 12, 22, 25, 30
Chaucer, Geoffrey, 9
Chester, 9
Chesterfield, 59
chisels, 13, 15
china clay, 39
clay, 8, 39, 40, 41
Claywheels Forge, 79, 81
Cleveland, 14
Cocker Brothers, 54, 70, 93
cogging mill, 86
coke, 13, 37, 44
coke breeze, 39
Cologne, 11
Company of Cutlers in Hallamshire, 9
Crowley, Ambrose, 11
crozzle, 106
crucibles, 8, 13, 39, 40, 41, 42

Cutlers' ordinances, 9
cutlery making, 9, 93
Cyclops Works, 60, 61, 62, 85, 89, 90

Davy Brothers, 72, 92
Derwent, River, 29
Derwencote, 11, 29
Don, River, 8, 12, 15, 64, 69, 72
Doncaster, 9, 12
Doncaster, Daniel, 26, 34, 42
dozzle, 47

Ecclesfield, 9, 93
Eckington, 9
edge tools, 15, 94
Ersta Steelworks, Sweden, 36
Eyre, Ward & Co, 62

Faraday, Michael, 53
fettling, 76
files, 8, 15, 96, 97, 102
finery, 10
Firth, Thomas, 13, 22, 30, 32, 56, 58, 59, 60, 79, 81
Firth-Brown Ltd, 47
Fischer, Georg, 53
forks, 95
Fox, Samuel, 71
Furness district, 14
Furnival Street, 54

ganister, 8
Gibralter Street, 26
Gloucester, 9
Goole, 12
Gosling, R., 16
Greaves, William, 62
grinding, 94, 98, 99
grinding wheels, 9, 18, 20
grog, 39

Hadfields Ltd, 50
haematite ore, 14
Hall, William, 97
Handsworth, 36
helve hammer, 11, 79
Hobson, Seaman & Co, 86, 96, 97
Holbrook, 9
Holly Street Furnaces, 30
Holmes Furnaces, 28, 29, 54, 74
Howarth, James, & Co, 94
Hoyle Street, 96
Hoyle Street Furnace, 26, 28
Humber, River, 12
Huntsman, Benjamin, 8, 13, 30, 36, 39, 41, 45, 48, 56, 93
hydraulic press, 15, 84

ingots, 8, 13, 14, 15, 46, 47, 48, 49, 54, 73, 74, 78, 81, 86
ingot moulds, 13, 14, 37, 38, 45, 46, 47, 48, 73, 78, 109
iron:
    Best Yorkshire, 35
    Dannemora, 11, 30, 35

Double Bullet, 34
GL, 35
Hoop L, 34
Spanish, 11
Swedish, 11, 30, 34, 69
Swedish White, 42, 69
Steinbuck, 35
W and Crown, 35
Warner's Carbonising, 42
iron ore, 8, 10
ironmaking, 9, 10

Jars, Gabriel, 25, 96
Jessop, William, & Sons, 13, 32, 39, 40, 42, 48, 50, 52, 64, 66, 67, 79, 86
Jessop Street, 64
Jowitt, Thos, & Sons, 2

Killamarsh Steel Works, 54
Kimberworth, 9
Kirkstead Abbey, 9, 10
knives:
    Kitchen, 102
    pocket, 104

ladle, 73, 76, 77, 78
Le Play, Prof F., 12, 22, 38
Lee, Arthur, & Sons, 93
Lescar, 9
Leufsta Forge, Sweden, 34
Leveson Street, 72
Lincolnshire, 14
Little Sheffield, 9
London, 9, 11
Loxley, 8

Machen, Miller & Machen, 69
machine shops, 15, 91, 92
Marsh Brothers, 46, 54
Masbrough, 12, 28, 29, 54, 74
Millsands, 9, 67
Millsands Works, 53
Millstone Grit, 9, 20
Moss & Gamble, 26

Nasmyth, James, 15, 83
Navigation Works, 54, 70
Naylor, Vickers & Co, 13, 49, 50, 67
New Sheffield, 85
Newburn on Tyne, 85
Newcastle, 8, 11, 29
Norfolk Works, 22, 56, 58, 60
Northamptonshire, 14
Norton, 9
Nursery Street, 70

Osborn, Samuel & Co, 46
Osterby Forge, Sweden, 34
Oughtibridge, Thos, 10, 17
Oxford, 9

packhorse, 12
Paris Exhibition, 62
Park Iron Works, 72, 92
Park Steel Works, 67

111

Parker, John, 9
Parkgate Works, 74, 76, 77, 78, 86, 87
pen nibs, 13
Penns Forge, 54
Perkins, 10
pig iron, 10, 74
pit coal, 8
Pittsburgh, 39
plate mills, 15, 89, 90
Pond Steelworks, 45, 54
Porter, River, 8
Prince of Wales, 49
puddled steel, 59, 60
puddling process, 14, 59, 60, 62, 74

railways:
   Great Central, 72
   Midland, 72
   Sheffield Manchester & Lincolnshire, 18, 20, 28, 62
   Sheffield & Rotherham, 13, 28, 58, 60, 61, 62, 69
razors, 13, 15
Rivelin, River, 8
River Don Works, 13, 49, 67, 69, 84, 89, 91
Robsahm, Ludwig, 36
rod mills, 86
Rodgers, Joseph, 93
rolling mills, 15, 85, 86, 87, 89, 90
Roman ironworking, 9, 10
Rother, River, 8
Rotherham, 9, 10, 12, 13, 28, 29, 54, 74

Salmon Pastures, 18
San Francisco, 50
Sanderson, Brothers, 53, 82
Sanderson, Charles, 53
Sanderson, Naylor & Co, 53
Savile Street, 54, 56, 58, 60, 61
saws, 8
scissors, 15, 104
Scotia Steel Works, 2
scythe making, 9, 15
Shah of Persia, 74
Sheaf, River, 8
Sheaf Works, 20, 62
shears, 95
sheep shears, 98, 101
sheet mills, 87
Sheffield:
   Catalogue, 101-5
   Park, 11
   plan, 1736, 16

Prospect, 17, 18
thwytel, 9
   Trades Historical Society, 108
   Views, 18, 20
Shepherd Wheel, 108
Shiregreen, 93
Shore, Samuel, 10
shovels, 95, 100
Siemens, William, 14
slashers, 101
Soho Rolling Mills, 67
South Yorkshire Coalfield, 8
spades, 94, 100
Spear & Jackson, 56, 60
Spencer, John, 85
spiegeleisen, 42
Stampel Bok, 34
Stannington, 39
steam power, 15, 82, 83, 85, 86
steel:
   acid, 14
   alloy, 10, 53
   American, 14
   armour plate, 14, 89, 90
   basic, 14, 74, 76
   Bessemer, 8, 14, 73, 74
   blister, 8, 10, 22, 28 (see also cementation steel)
   bulk, 8, 14
   carbon, 8
   cast, see crucible steel
   cementation, 8, 9, 10, 11, 12, 17, 18, 20, 22, 24, 25, 26, 28, 29, 30, 39, 42, 53, 54, 58, 60, 61, 62, 64, 66
   crucible, 8, 12, 13, 14, 15, 28, 36, 37, 38, 39, 40, 41, 42, 44, 45, 46, 47, 48, 49, 50, 52, 53, 54, 60, 62, 66, 67, 69, 70, 71, 73, 78, 108, 109
   Cullen, 11
   double shear, 12, 13
   forging, 15, 79, 81, 82, 83, 84, 94, 95, 96
   German, 11
   gun barrel, 14
   heat treatment, 15
   imports, 8, 10
   open hearth, 8, 14, 74, 76, 77, 78
   phosphorus, 14, 76
   plate, 87
   rails, 14
   rolled rings, 90
   shear (single shear), 12, 98
   sheet, 13, 87
   shells, 14

   special, 8, 10, 14, 15
   stainless, 15, 89
   wire, 13, 70, 89
Steel, Peach & Tozer, 90
Stocksbridge, 71
Stourbridge, 10, 11, 25, 39
Stubs, Peter, 28, 54
Sutton Coldfield, 54

Tankersley Coal Seam, 10
Templeborough, 9
Thomas, Gilchrist, 14
Thorpe Hesley, 9
Thwaites & Carbutt, 83
tilt hammer, 15, 82, 83, 109
Tinsley, 12
Toledo Steel Works, 69
Tool Cabinet, 105
trade marks, 9, 34, 35
transport, 9
Trent, River, 12
tundish, 48, 50
turning tools, 8, 13, 15, 102
turnpike roads, 12

Vickers, Thomas, 69
Vickers, Sons & Co, 69
Vickers, Son & Maxim, 84
Victoria Station, 18, 20, 62

Wadsley, 10
Wadsley Bridge, 9, 79, 80
Wadsley Bridge Steel Works, 69
Walker and Booth, 12, 37
Walkley, 10
Wardsend, 70
Warrington, 28
water power, 8, 15, 79, 81, 82, 85, 108, 109
Webster, Joseph, 54
Webster & Horsfall, 54
West Street Works, 53
wheelswarf, 12, 25
Whittington Works, 59
Wicker Station, 13, 61
Wicker Wheel, 16
Wisewood, 9
wortle plate, 93
Wortley Forge, 11, 79, 85
Wortley Wire Mill, 93
wrought iron, 10, 14, 60

TS
304
G7
B37

JUN 27 1978